CURIOSITY

ALSO BY ROD PYLE

Destination Mars: New Explorations of the Red Planet

CURIOSITY

An Inside Look at the MARS ROVER MISSION
and the PEOPLE WHO MADE IT HAPPEN

ROD PYLE

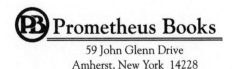

Prometheus Books

59 John Glenn Drive
Amherst, New York 14228

Published 2014 by Prometheus Books

Cover images © Media Bakery
Cover design by Nicole Sommer-Lecht

Inquiries should be addressed to
Prometheus Books
59 John Glenn Drive
Amherst, New York 14228
VOICE: 716–691–0133
FAX: 716–691–0137
WWW.PROMETHEUSBOOKS.COM

18 17 16 15 14   5 4 3 2 1

Library of Congress Cataloging-in-Publication Data

Pyle, Rod, author.
    Curiosity : an inside look at the Mars rover mission and the people who made it happen / by Rod Pyle.
        pages cm
    Includes bibliographical references and index.
    ISBN 978-1-61614-933-8 (paperback) — ISBN 978-1-61614-934-5 (ebook)
    1. Roving vehicles (Astronautics) 2. Curiosity (Spacecraft) 3. Mars (Planet)—Exploration.
4. United States. National Aeronautics and Space Administration. I. Title.

TL799.M3P95 2014
559.9'23—dc23

2014007558

Printed in the United States of America

To my son, Connor:
The future is yours,
use it wisely.

CONTENTS

ACKNOWLEDGMENTS

Thanks are due to countless people who generously gave of their time and energies to assist in the completion of this book. The bulk of these people work at JPL, either on staff or as contractors, and rarely get the thanks or recognition they deserve.

First and foremost: my deepest thanks to the selfless Guy Webster, as fine a person as you are likely to meet during your brief time on our planet. He is the media's point man for JPL's Mars program, and gives and gives without complaint or hesitation, which is rare in his area of endeavor. And—he knows his stuff; Mars is no stranger. I simply cannot thank you enough.

John Grotzinger, who was kind enough to survive my many questions, most of which are well below his pay grade. John made time for me on numerous occasions when I'm sure he had better things to do (like exploring Mars, for example). John, I am in your debt for the many hours you spent educating me on things I did not know I needed to know, and making it fun.

Rob Manning, a fantastic guy, gave me more time than I could have asked for and did so with copious good grace. He is a font of information presented in a way that is as enjoyable as it is informative, and the fact that he enjoys it so much means that all of us do too.

Very special thanks to: Ashwin Vasavada, Dan Limonadi, Joy Crisp, Justin Maki, Mike Malin, Ken Edgett, Scott McLennon, Lauren DeFlores, Vandi Tompkins, Doug Ming, Brian Cooper, Rebecca Williams, Al Chen, Steve Squyres, Chris McKay and Jakob van Zyl. You all provided time for interviews in schedules that were already jammed to the hilt. My thanks.

To Buzz Aldrin, who made time for questions about this story as well as others; you are a founder of the Space Age, and one of the few key players who continuously strives to push open the frontiers. Mars owes you much.

Robert Zubrin, whose work underlies so much of manned Mars planning now it's hard to find the borders of his involvement. Thanks for being a selfless champion of the red planet. Alexandra Hall of the Google Lunar XPrize: you will first win the moon, then Mars. Carry on.

Steven Dick, former NASA chief historian and now astrobiology chair for the Library of Congress, and Roger Launius of the Smithsonian NASM, thanks as always. Leonard David, space journalist and fellow niche author: thanks for being there.

Steven L. Mitchell of Prometheus Books: your continued support of planetary exploration and the sciences at large are a testament to the integrity and high standards of Prometheus. We are all better for your efforts. And to the rest of the Prometheus Books staff: Catherine Roberts-Abel, Melissa Raé Shofner, Bruce Carle, Nicole Sommer-Lecht, Jade Zora Scibilia, Mariel Bard, Meghan Quinn, and those who worked behind the scenes—thank you.

John Willig, ever-present agent and supportive friend, who wrangled the details and made it look easy as he always does. Deepest thanks once again.

Blaine Baggett runs a tight ship at JPL and it shows. The communication division works wonders with limited resources. Veronica McGregor, Jane Platt, Jia-Rui Cook and D. C. Agle make this sort of thing possible.

To the rest of JPL's able PR staff: Elena Mejia, Mark Petrovich, Daniel Goods, Erik Conway, Scott Hulme, and John Beck-Hoffman—thanks for responding to my continuous (and probably irritating) requests for media materials. There is simply no way for an author to approach such as task without your talents and support.

Lawren Markle and Brian Bell of Caltech's media relations: thanks for your accommodation. Lawren has since moved on, but may well have saved me from becoming a permanent part of Death Valley.

Janice Alvarez provided tireless transcriptions on time and with remarkably few errors or omissions considering the complexity of the work—and while delivering a new baby, no less. I have no idea how you did it.

Thanks to my pals from the Griffith Observatory days who generously donated their time to JPL during the Curiosity landing: Jim Somers and John Sepikas. You made the landing a whole lot easier for the press; they will never know what you did for them. Glenn Miller helps you stay the course. And Bob Brooks, who is at JPL every day—Mars would not be the same without you.

Thanks to Ken Kramer, Sherry Clark, and Scott Forbes, for being there. To my son, Connor, who endured countless hours of silence and separation (do teenagers even notice?) while I operated in missile-silo mode to write: thanks for being understanding.

Sherry Clark, you are a pillar of strength and support. Gloria Lum, your contribution to this process is too much to write. Thanks.

To mom and dad: thanks for being understanding of a sometimes-puzzling son.

To the readers: thanks for being understanding of a sometimes-puzzling adult. I love writing books, and your support is the greatest treasure any author can ask for.

AUTHOR'S NOTE

Any book such as this requires a vast amount of research and cooperation with the National Aeronautics and Space Administration (NASA), the Jet Propulsion Laboratory (JPL), Caltech, and other appendages of the government/academic space-exploration enterprise. Of course, the richest resources are the people within, and the larger NASA entity is filled with wonderful and helpful ones who have vast experience and long memories. In particular, the exploration of Mars is an undertaking that inspires great passion in the participants, fostering deeply moving and fascinating conversations.

I interviewed dozens of participants in the Mars Science Laboratory (MSL) mission for this book, and it draws upon their statements and memories liberally. In all cases they demonstrated an acute memory of entry, descent, and landing akin to people's memories of where they were when Apollo 11 landed on the moon, or for the younger generation, that awful day when the space shuttle *Columbia* disintegrated during reentry. Other memories and recollections of duties and actions taken since landing, during the primary mission, were also fresh. For the scientists, their summaries of their particular areas of interest were detailed and exact. Likewise for the engineers; their recall of their portions of the mission and the technologies within were acute and complete. These recorded interviews were later transcribed into about three hundred pages of text for reference.

For the recounting of specific statements and events not covered within interviews (or requiring backup), I relied on two primary sources. The first, where applicable, were the vast amounts of archival video footage of various phases of the MSL mission which are available online both through publicly accessible NASA websites and by request via JPL's servers. One may also order footage from JPL's vendor if desired. Thousands of still photos back up this visual data, too. The second source was the statements of other participants in the mission themselves, and in cases where there was any mismatch or doubt, I attempted to find a third source to support individual claims.

In absence of other recorded data, Guy Webster, JPL's point man for Mars, has a vast institutional memory and was helpful when other methods failed. He is a rich resource.

For older missions and histories, NASA has massive archives spread across the web and in various physical locations—specific to planetary exploration, JPL, Caltech, and Nils Bohr Institute archives can be particularly helpful.

All this said, one occasionally finds an error or misquote that has been propagated across time and space—something printed twenty-five or thirty years ago, then erroneously quoted in another book or article, *then* archived among a half dozen or more web portals, can become very thorny to unravel, especially when (as sometimes occurs) the primary reference may have gone missing. Fortunately, MSL is a young-enough mission that this is not yet a concern. In fact, part of NASA/JPL-Caltech's challenge with this mission is to preserve and archive events and documents as they happen, with very little budget to do so.

The vast bulk of the imagery in the book comes courtesy of NASA/JPL-Caltech. Their online image archives are superlative, and in most cases the accompanying captions make clear the phase of the mission related to the photo at hand. A few other images were sourced from stock agencies or Creative Commons sources. Still others come from my personal collection.

Review and some fact-checking was performed by JPL and in some cases specific mission participants. Any remaining mistakes are my own.

CHAPTER 1

DEATH VALLEY DAYS

Vultures cut lazy circles in the sky above me, patiently awaiting my certain demise. Dropping my gaze, I could see the heat shimmering off the mountainside. It was well over one hundred degrees and the last of the water was long gone. I was close to forty miles from the nearest road and the situation looked bleaker than the late dinner seating on the *Titanic*.

What did this potentially life-threatening situation have to do with Curiosity on Mars?

Then the image of Lawren Markle, the dapper (and sweatless, in stark contrast to my drenched shirt) PR rep from Caltech, swam into focus before my parched and crispy eyes.

"You doing OK?" he said with a look of genuine and entirely reasonable concern, given my condition. "Sure," I croaked. "I'm . . . good. . . ." It was then that the death fantasy evaporated and I was back in an equally grim state—atop a barren, rocky peak in Death Valley with a group of science journalists and our Edmund Hillary–esque geological guide, Caltech's Dr. John Grotzinger.

I looked from Grotzinger, who was holding forth on a geological premise that was not quite soaking in, to Lawren, who had refocused his attention to Grotzinger, where it belonged. Having Grotzinger as our tour guide in this garden spot of Death Valley was a true privilege. If I survived it, I would be very grateful in a week or so, once I was able to rehydrate.

Reality hit: we were probably a mile from the road, it was about eighty-four degrees, and we had been hiking for about twenty minutes. I was wiped out. This was not promising to be my best day.

The invitation I received in my e-mail a week before had sounded exciting and bore the return address of Caltech. I was invited to join Dr. John Grotzinger, mission scientist (a.k.a. Big Kahuna) for the Mars Science Laboratory, as he hosted a two-day press junket into Death Valley to discuss some of the formations that informed earthbound geologists about what they might find on parts of Mars. It all looked like good fun, and there would be only a handful of us, maybe twelve science writers. I would be the sole outside video guy; JPL's John Beck-Hoffman would shoot for their media operation.

At the end of the note was added, "Wear sturdy shoes, as we will do a bit of walking." Oddly, I missed the hidden meaning even though I had taken a string of geology classes in college. I should have remembered that this was geologist's code for "wear steel-toed boots, as we are likely to run into rattlers and possibly even desert snarks as we hike over hill and dale looking for the perfect Precambrian outcrop" . . . or in this case, as we hike about twelve miles, mostly vertical, in the hottest and driest place in the continental United States.

What may have thrown me off was that Grotzinger is not the old-breed of geologist I knew from my increasingly distant college days. He wore no suspenders, no plaid flannel shirt, and did not smoke a pipe. He had no gold-miner's belly hanging over a turquoise-buckled beltline. He is, in fact, tall, slim, irritatingly fit, and tan. His weathered good looks would be welcomed into the Explorer's Club. While his explanations can be challenging to follow, he is patient and graceful with the laypeople. He can be folksy and technical in the same sentence, so clearly the geologist's pedigree was there.

We traversed many roads and junctions on our way into the high desert, and the map showed many nearby avenues with such charming (and informative) names like Talc Road, Sulfate Road, and Cactus Flats Road. A few hours out of LA, we reached our destination: Shoshone, California, with a population of thirty-one. Our group almost doubled it.

As we checked in at the cinder-block penitentiary that was to serve as our motel for the night, an optimistically named "inn," Grotzinger wandered over to one of the hulking, chalk-white Caltech SUVs that had transported us there. By the time I got my camera equipment and other packings stowed in the room, he was completing elaborate dry-marker murals on the rear passenger-side door of one of these overwrought grocery haulers. Big murals. Murals of complicated rock bedding, uplift, and other geological delights. And on the front door, he had drawn surprisingly artistic renderings of very steep Alpine-looking mountains, which I hoped were metaphorical. As I watched, and thought back to my own college geology field trips, I wondered if any of his compatriots had ever swapped out a permanent marker in the SUV drawing kit just for laughs. I know I would have. "Ha ha, that was funny. Here's the turpentine . . . oh, that takes off the car's paint too, now doesn't it . . ."

As we waited for the formal talk to begin, a three-foot-long and very red snake worked its way across the roadway and for some reason (probably seeking the same relieving shade that I would soon pine for) decided to work its way through the spokes of the truck's wheel. Our group tittered nervously. Most of us were now regretting the light athletic shoes we had chosen to wear.

We moved to a fire ring nearby (why a fire ring in Death Valley, you might ask? To keep the sidewinders and rabid coyotes away at night, of course), and Grotzinger began discussing the day ahead. I missed most of it because I don't shoot video out on location all that often and was fiddling with my camera gear and cleaning the sand off the parts I had already dropped twice. But I had it running by the time he got to the most immediately relevant part of the talk, moving back to the mural on the Suburban. "Then we'll end up at this formation, here in the foothills. It's the Glimmenshorp Finkleheimer formation of preglandular hermphratite . . ." (this was not exactly what was said, but it's how it registered in my overheated brain).

I was still rolling video, so I meekly asked, "John, for *my* audience would you mind giving me the eighth-grade, *Weekly Reader* version of the story?" He paused for a moment, looking at the ground. I thought perhaps I had committed some kind of faux pas, but I think he was merely gathering his thoughts. At least I hope so.

"Um, OK. So what these structures represent is a feature called stromatolites, that we interpret on Earth to represent the interplay between microbial mats that are living on the seafloor and sediment that comes in and interacts with the mat. The sediment is ultimately why it becomes a rock. If it's just left as a microbial mat, the organics will decay and there will be no record of that, but by having the sediment come in, it can get cemented and turn into a rock. Then we see these features in the rock record and infer that there were once microbes living on the seafloor that were forming these structures."

He paused for a moment to make sure I was getting it. At least the camera was. He continued: "What it has to do with the MSL [Mars Science Laboratory] mission is that it represents an end member of a feature that, if we saw it on Mars, we would stop and definitely study the outcrop. Immediately there would be a lot of discussion, because even on Earth we cannot take this rock and bring it back to the lab and prove to you unequivocally that this feature represents evidence of life on the early Earth." As we will learn later, and I promise it will be simpler, on Earth these stromatolite-like features *can* be caused by living things, or they can mimicked by processes having nothing to do with life. On Mars it would be (a) a wonderful shock and (b) a guaranteed conundrum. "[On Earth] if you put in the context of other features that occur along with it, it's a perfectly reasonable suggestion that these were microbial mats on a seafloor. But independent of that, in rocks of this age, we often see microfossils. I would say that if we ever found something like this with MSL, we would stop and study it, and it might be a really good place to come back to someday in the future and do a sample return to Earth." Not to mention a press sensation and a tenfold increase in planetary-exploration budgets. But I digress.

He stopped, flashed what felt like a sympathetic smile, and it was time to move on with the adult portion of our program. I envied the grad-student assistants present that they could understand the version that was to come.

Soon we headed out to the first stop on the side of the scenic Western Talc Mine Road. Once disembarked, Grotzinger began with a sweeping gesture toward some distant hills. "See those white areas over there? They are due to this rock," He gestured to a chart he had prepared, "which is shown to have a cross-cutting relationship where it comes up and goes into the Crystal Springs formation there . . ." Pointing to a specific spot (that looked to this layman like more gray hillside), "It's a basaltic composition rock, and that's a Mars kind of rock. It's not a granite; it's got olivine and pyroxene in it, or once had it, and it intruded into the Crystal Spring adjacent to some carbonate rocks . . . and it has altered the rock. You'll hear that from the Mars guys, especially the mineralogists, talking about alteration. Well, that white rock is a result of alteration, it's a mineral called talc."

I felt a moment of pride, for I knew what talc was. It comes as a powder in bottles and lives in bathrooms.

"We're going to go right up there to see the talc and the carbonates that it's intruding into, so you will see the source of the heat which is the basaltic rock. It intruded in, and then it heated up the sediments which were wet, and it made talc, which is a hydrated magnesium silicate mineral, kind of like a clay."

Aha! Understanding crept into my brain. If Curiosity found clays on Mars, that would be evidence of water, and there were many complex processes under which this alteration into clays might take place. This one had involved intense heating.

Fig. 1.1. STROMATOLITES: This is one form (of many) that stromatolites can take. They can be caused by either biological or nonliving activities, and it can be difficult to discern one from the other. Still, were something like this found on Mars, it would grab the attention of the geologists. *Image from Mark A. Wilson (Wilson44691).*

Overall, this was a dense subject, somewhat above my pay grade, but clearly Grotzinger loved the topic and while it may have been a challenge to talk at our level, his enthusiasm was highly infectious.

We loaded up the trucks and headed out to some barren, rocky mountains in the lowest, hottest place of America.

Less than an hour later, we were in the foothills heading to the first hike of the day. Leading up to this, the sights had been within a few dozen yards of the highway. This had suited me fine, what with my twelve-pound camera and eighteen-pound aluminum tripod (the lighter, carbon-fiber ones are expensive . . . I pride myself on equipment frugality and never understood those guys who spent thousands on carbon-fiber tripods—I soon would). It was early and I had already sweated out a few quarts into my designer tropical shirt, vented at the back as a fashion statement.

As we unpacked for the hike, I took stock of my companions. There were about a dozen of us, including Mike Wall, my compatriot from Space.com. He was writing a story and I was shooting one. I would soon admire his choice of a pen as an instrument of communication over my camera. Others were from the *Wall Street Journal*, *Reuters*, *New York Times*, *Washington Post*, *New Scientist*, and the *LA Times*. And then there was Beck-Hoffman, the JPL camera guy who was annoyingly fit and nimble (and who would soon be known as the *auteur* behind the "7 Minutes of Terror" video that was downloaded millions of times). He was carrying a large, shoulder-mounted camcorder. It would slow him down far less than my smaller rig would me.

We parked on a mesa and headed off, following Grotzinger toward a series of hills. They didn't look like much from the starting point. But by the time we got to the bottom of them, they were impressive indeed. Funny how these things magnify in intimidation when you get closer—the desert compresses everything from a distance. The group made its way up, each at his or her own pace but in little clots of people—and then there was me. To be fair, I was carrying the camera in one hand and the tripod over the opposite shoulder, trying to balance myself as we climbed the rough, bouldery hillside. At least that's the excuse I was using. Within moments, I was out of breath, panting like a dying asthmatic, and falling behind. I looked from side to side—there were mounds here, surely shallow graves of journalists who had gone before me. But I stuck with the program.

When I reached the top, the lecture had begun and the other journalists were taking copious notes. I set up my camera with slippery, sweaty hands, and once the ringing in my ears abated, which for a moment I thought might be the signs of a mild stroke, I listened to Grotzinger speak on the formations at hand.

No sooner had I caught my breath than we were on our way farther up the hill. To the top. Which looked very far away. I shouldered the camera rig and followed at an ever-increasing distance, Lawren shooting worried (and deservedly so) glances my way from time to time. It's nice to be cared about when you're in extremis. On the other hand, I guess it would have been bad publicity for Caltech to host the death of overweight, out-of-shape, late-middle-aged journalist.

For the rest of the day we climbed, descended, and climbed again. The rocks were sharp and nasty. Grotzinger literally scampered hither and yon and most of the journalists were keeping pace, albeit with an increasing gap between themselves and Grotzinger. And then there was the gap between them and me. Leave me, save yourselves.

Fig. 1.2. WHY THEY CALL IT DEATH VALLEY: This was not the region we visited, but it's close enough to give you an idea of how the place got its name. Fascinating, lovely . . . and on the wrong day, deadly. Summer temperatures can top 130°F. *Image from iofoto.*

We came to another mountaintop (OK, it was a hilltop but give me some dignity), and Grotzinger waited for the group to settle. I approached as he began his talk. He was not even sweating. "If you look carefully enough, you will see that these things all go in the same direction. So if you walk up a bit you'll see more of them, then we'll go around the corner where you can see a really big one, about a meter in diameter. Have a look, then we can talk some more about it." He walked off to chat with another husky, bearded fellow (he could pass as a younger version of me, without the cardiac issues), Ken Edgett. Ken is a geologist who works for

Malin Space Science Systems (MSSS), makers of the main imaging cameras on Curiosity (and imaging systems of other spacecraft that have orbited Mars). He wasn't sweating either.

The landscape was pretty in a bleak way, but I had missed the actual subject of the stop. I resolved to be more prompt to the next one, even if it meant an early demise.

The remainder of that day and the next followed in a similar pattern, and a night of deep sleep did little to lessen the impact of the second day's calisthenics. It was informative but challenging. At the end of the second afternoon, I pigeonholed Mike Wall and asked him to summarize the trip. He has a doctorate after all (though in an unrelated field) and is the senior writer for Space.com.

He summarized the excursion succinctly: "John walked us through what field geologists do, how you can make sense of all the different rock layers, what they mean, how they were deposited, and so forth. It was a great exercise. Couched in this is what Curiosity will be doing on Mars and how we can try to make sense of the sort of rocks and formations that we may find in Gale Crater."

He thought for a moment and continued: "It's a process, it's all context and getting to know the environment. We learned that it's harder than you think it is, that you really need to know the rocks. It's going to be hard for a robot to do that on a different planet, but MSL has a great support team, over four hundred scientists who are working on the project. They will interpret all this incredible data that the rover will gather. They are very smart people and will figure it out."

Looking at my sweat-soaked shirt, he smirked and added, "It's windy, but not even one hundred yet, maybe ninety-five degrees. Easy."

I thanked him and returned to the air-conditioned Suburban. We headed back to civilization to witness the beginning of the most ambitious and incredible mission to the surface of another world ever attempted.

CHAPTER 2

TANGO DELTA: TOUCHDOWN

A s you have probably surmised, John Grotzinger is not normally a nervous man. A career geologist and professor at the California Institute of Technology (Caltech), he seems equally at home running a huge spaceflight project as he is in front of a class of anxious undergraduates. He jokes easily, and his explanations can be disarmingly folksy but can also turn deadly serious when explaining the relevance of contact metamorphism in a rock. Both are useful traits in his area of endeavor.

But tonight is different. Tonight the stakes are high indeed. Tonight, $2.5 billion of Mars rover, the aptly named Curiosity, will either land in Gale Crater to begin its twenty-four-month (one Martian year) primary mission, or pinwheel into costly wreckage across a couple hundred acres of Martian desert. Tonight will be either a validation of a decade of planning, designing, building, and launching of a wonderful era of exploration, or the likely end of America's Mars exploration program.

So tonight, Grotzinger seems as nervous as a cat on the freeway.

You would never guess it just by looking, of course; geologists don't roll like that. It can be easier to gauge the feelings of his associates in mission control at the Jet Propulsion Laboratory. Dozens of engineers, controllers, scientists, and other highly proficient people surround him in the final moments of the flying phase of Curiosity's mission, and some display their feelings a bit more directly. Perhaps it is because each of them has been intensely focused on a smaller part of the mission— parachute deployment, entry and guidance, pyrotechnics, software design, or one of a dozen other specialized areas. Grotzinger, as the principal scientist of the mission, must be conversant in each specialization of MSL yet also possess a more global perspective, that is, to oversee the mission once on the ground. Tonight there is not much that he can do except wait and hope.

A few consoles forward sits a man whose face is known to Mars enthusiasts worldwide but whose name may not be. Middle-aged and graying, he is nonetheless clearly still the smiling, ever-cheerful man we saw flashed so memorably across the Internet, fist-pumping the air when Mars Pathfinder landed successfully in 1997. He was thinner and darker-haired then, but the effervescent energy is still there. Rob Manning may have matured, but he is once again the chief engineer in charge of a Mars machine, and there is nowhere he would rather be. It shows.

Rob is not a professor, nor does he hold any official positions outside JPL. He has been the main machine guy for every Mars rover to date—that is his raison d'être. He is also father to two precocious daughters and plays jazz trumpet gigs in his scant spare time. Back at work, after sweating out the pioneering Pathfinder mission in 1997 and the design and landings of both Mars Exploration Rovers, Spirit and Opportunity, in 2004, he is at the crescendo of his career with Curiosity. It's been a long haul, though, and the endless hours of meetings and checkouts and visits to the cape have taken a toll. He will rest when the rover is on the ground—in one piece or in hundreds. For now his normally cheerful face shows intense concentration as he follows the data coming back from Mars. So far, Curiosity, as much a third child to him as any machine could be, is doing fine.

Down near the front sits a man whose face—or, to be more precise, whose hair—will soon be iconic. Bobak Ferdowsi sports a wild, multicolored Mohawk hairstyle, a personal statement of celebration for this momentous mission. It's like nothing we have ever seen in mission control, and within hours the Internet is alive with tweets and posts about this new phenomenon. Bobak, in his late thirties, is a handsome young man who has worked for almost a decade on MSL's launch, cruise, and approach phases. His role in these final moments of landing is largely one of quiet observation, watching data he knows are already fourteen minutes old, delayed by the huge distance between Earth and Mars. He summarized it well: "It's like taking the SAT [college entrance exam]—you've taken the test, you are done. Nobody will tell you the score right then, so you have to wait. You're nervous . . . something happening now could change the outcome of your life." And he couldn't be more correct, except that his life will change in ways far different than he expects, including a couple of hundred marriage proposals via Twitter and a future visit to the White House to meet the president and First Lady. But that is some time off, and at the moment he is fixated on the screen in front of him.

But not everyone involved with the mission can fit in mission control. A building away, Ashwin Vasavada sits, also intently eyeing a computer screen. He is, along with

Joy Crisp, a deputy project scientist on MSL. He is also a spectator at this point. Curiosity carries experiments that he and Joy have helped to shepherd through the process of construction, testing, installation, and launch, as they oversee the larger science team behind the MSL mission. "I feel like I'm enabling almost 480 scientists and a spectacularly good mission with a lot of scientific integrity," he comments. "I like to put my heart and soul into making sure we do good science." Now, like Grotzinger, all he can do is watch, a helpless captive to events far away, above another world, as the much-anticipated entry, descent, and landing (or EDL) sequence begins.

Vandi Tompkins, slated to be one of a small group of rover drivers and programmers for Curiosity, is also waiting out the landing as an observer. As an engineer, she has ultimate faith in the machine and her comrades who designed it. As a PhD in advanced robotics, she also has a deep attachment to the machine in a way that few would understand. And as a woman who came to the United States from India to make her way into the stratified world of planetary exploration, she can sometimes scarcely believe her good fortune to be here, on this program, at this moment. The next few minutes may well determine her fate for the next decade.

And this is it—the by now well-known seven minutes of terror, made famous by the video of the same name that went viral a couple of months earlier. On cue, MSL spacecraft plunges into the thin Martian atmosphere after a largely uneventful journey lasting nine months. Much like war, robotic spaceflight is often characterized as long stretches of boredom followed by moments of intense terror. The spacecraft will automatically go through the intricate ballet of landing in a small target area on a planet far, far away. When it is over, the signals describing success or failure will still be making their way back to Earth at the speed of light; during landing, MSL (and the Curiosity rover ensconced within) is completely on its own. Grotzinger, Manning, Ferdowsi, Vasavada, Tompkins, and thousands of others will still be staring at computer screens, continuing to count the seconds and ticking off the milestones as if events were still unfolding. In short, regardless of Curiosity's fate, people in Pasadena and all over the world will continue watching and waiting for the signal that will free them to breathe again.

High in the skies of Mars, following a roughly equatorial trajectory, MSL hurtles toward the Martian surface. The onboard computer, a radiation-hardened version of an early-2000s Macintosh PowerPC chip, is processing incoming data like mad and adjusting flight parameters to match. A firing of a thruster here, a guidance adjustment there. There's not much time left to correct anything, though, for the guided portion of entry—where MSL glides across the skies on its heat shield, adjusting course continuously—is just about over.

On cue, the huge parachute deploys, unfurling spectacularly above the spacecraft while it is still traveling at supersonic speed. The craft begins to slow and angle toward the ground. High above, in Martian orbit, one of JPL's other robotic explorers, the Mars Reconnaissance Orbiter, snaps a photo of a tiny spacecraft dangling from a parachute. That it captures the lander at all is a near miracle, for it was being aimed in advance purely by sophisticated calculations performed by JPL flight engineers. The resulting image was a remarkable bit of space-age orchestration.

Down, down Curiosity drops, until—while still high in the air and in an apparent contravention of logic—it separates from the parachute. But seconds later, powerful onboard rockets fire, further slowing the spacecraft.

At JPL, a room full of men and women, mostly in their twenties and thirties, monitor the hundreds of systems and subsystems critical to a successful landing. The old folks—the ones over fifty—are mostly in the back row. This is a young person's mission.

Adam Steltzner, the forty-nine-year old rock 'n' roller primarily responsible for getting Curiosity onto the surface of Mars safely, paces like a lion on the hunt behind a row of intent controllers. He's been likened to Elvis with a doctorate in engineering, and now he monitors the consoles and the big picture. His pace quickens as the displays show MSL getting closer to Mars.

Events are occurring rapidly. Data from multiple onboard radars are guiding the machine to a landing inside Gale Crater, and within that into a landing zone called a "landing ellipse" due to the shape. It is just over five miles in width, and its long axis is twelve miles. It is the most ambitiously accurate landing attempt yet, and from what the folks back home can see, as smooth as any to date. The rocket pack/descent stage and its associated winching mechanism, collectively dubbed "sky crane," is working perfectly. In the final phase of what looks like a landing cycle designed by Rube Goldberg, or possibly Wile E. Coyote, the descent stage slows to a walking pace and sky crane begins to winch the Curiosity rover down via four cords, each about sixty feet long. It is by far the most complex interplanetary robotic undertaking in history.

Manning is at one of the rearward consoles, not quite with the old guys in the back but not with the youngsters at the front either. He's watching the system he led his team to create—not just sky crane but the whole landing system—and probably also thinking about the last-minute fixes he applied to Curiosity while it was already bolted atop the rocket, ready to launch. It was a close thing.

Al Chen is on the console as the operations lead for this critical phase of the mission and also the voice of MSL tonight. He is in his thirties, is married with three kids, and is normally a pretty unassuming guy. MSL has thrust him into the limelight, and his announcements come over the speakers in hushed, almost-unbelieving tones: "Sky crane deploying." It's JPL shorthand for "Holy s—! This damn thing works!" A collectively held breath releases and some scattered cheering issues forth, followed by applause. Then, a few moments later: "Touchdown confirmed—we're safe on Mars!"

The room goes nuts as controllers, scientists, engineers, and other associated JPL'ers erupt in heartfelt cheers. In a nearby building, inside the press room, normally stoic and hardened reporters from the major TV networks, newspapers, and periodicals just lose it. It is true pandemonium. Curiosity has arrived . . . and done so to perfection.

On Mars, MSL's rocket pack, comprising the engines and the navigational unit that brought the rover to the surface, has long since detached and flown off to crash a few miles distant. After a muffled thump, silence returned to Mars.

It will later be concluded that the rover alighted about 1.5 miles from point zero, well within the landing ellipse. It's a true pinpoint landing, as close to Mount Sharp, their primary objective, as anyone dared to hope.

Fig. 2.1. THE VOICE OF EDL: Al Chen on the console shortly before touchdown, narrating the entry, descent, and landing phases like a play-by-play. *Image from NASA/JPL-Caltech.*

Once the machine is secured, a jubilant EDL team bursts forth from mission control, heading to the press room (in more normal times, JPL's auditorium), wearing matching powder-blue polo shirts and chanting "Eee-Dee-El! Eee-Dee-El!" in triumph. It is a rare moment of collective joy and outright, raw emotion for these normally reserved people as they literally dance across the university-like quad. And it is well deserved.

Grotzinger smiles broadly, hugs a few people, pumps even more hands, and prepares for the press conference soon to follow. Rob Manning is mobbed by well-wishers, smiles like the punch-drunk engineer he is (he has been at this for eight intense years and has not slept much for almost thirty-five hours), and heads off to a series of interviews. Joy and Vandi hug coworkers and then begin to think ahead. Unlike Grotzinger, they do not have the camera lights and questions of a press conference to distract their attention from the challenges that lie before them. Both return to thinking about the larger mission—the upcoming milestones needed to ensure that the rover can accomplish its primary objectives.

Bobak Ferdowsi takes a moment to check his smartphone and discovers that in the past few hours, he has become an Internet meme and instant sensation. But there is not time for that now; he has people to congratulate and a rendezvous with a pillow. Ashwin feels the glow begin to fade a bit. He is responsible for coordinating the moment-to-moment science activities of the mission, a massive job, and tomorrow things will get very busy regardless of how much—or how little—sleep he manages to get. Nonetheless, he will remember the landing as one of the high points of his life.

Hundreds of others on-lab, and hundreds more offsite, pop champagne corks and toast success tonight. They enjoy the moment, as well they should, for tomorrow engineers, scientists, and managers begin a multimonth grind known as Mars Time—their days will match Martian days, known as "sols," which add forty minutes to each twenty-four-hour Earth day. The world will soon become a surreal, time-shifted place to the bleary-eyed participants in Curiosity's mission. But tonight is for celebration.

On Mars, the dust has settled around the unmoving rover. A few clicks and whirs can be heard from the inside as the Hazcams pop open lens coverings and begin imaging the immediate surroundings. Other mechanisms restrained for landing free themselves, and heating systems power on for the cold Martian night ahead.

Curiosity has arrived, and the greatest adventure Mars has known is about to begin.

CHAPTER 3

A LONG ROAD TO A RED PLANET

Exploring Mars has been a passion of earthbound scientists for centuries. Few thought of actually *going* there until the mid-1800s, but notions of Mars as a planet where beings might exist date back nearly two hundred years earlier. Christiaan Huygens, a Dutch astronomer and cosmologist, published a book in 1698 that speculated about life on the red planet. This idea bloomed in the 1800s, but it would be much later that the idea of exploration of the planet with humans was taken seriously. The idea of using machines to do so came later still. Many early visions harkened back to the golden age of earthly exploration, which was not conducted by robotic machines—there weren't any to do so yet. These voyages were accomplished by *men* in square-rigged ships that traveled to distant and forbidding places to plant the flag of their nations in faraway lands (and often cause disease and destruction to rampage through the native inhabitants). Surely, the thought went, the exploration of Mars would be accomplished in a similar fashion—by men (probably military) in rocket ships.

While many over the centuries thought of traveling to other worlds, it was not until Mars was understood as a planet—that is, a *place*—that these notions gained any sophistication. But even very early observers with scientific grounding began to notice things about Mars that were telling. In the fourth century BCE, Aristotle noted that Mars was occulted by the moon—Mars clearly passed behind it. This led him to the conclusion that Mars was farther away from Earth than the moon was. While this may seem painfully obvious to us today, at the time little was known about *anything* in the night sky; certainly Mars had not yet been identified as a world.

The earliest observations of Mars centered on its color. Few objects in the night sky exhibited any hues other than white, and none were as ruddy as Mars. So it is not surprising that the planet became intimately identified with all things violent—war, famine, and death. The various mythic figures that Mars became associated with were murderous at worst and portended a bad day at best. It was the Hannibal Lecter of celestial bodies.

Ancient Babylon, Egypt, and China all had deities associated with Mars. To the Babylonians, Mars was associated with Nergal, a deity of fire and destruction. To the Egyptians, Mars was Horus the Red, associated at first with the harvest, but later with violence. China and other Asian cultures saw the planet as representing fire and all its unpleasant associations, and eventually with mayhem. You didn't want Mars (or Nergal, Horus, or even the combustible Asian version) dating your daughter.

The Greeks also worshipped Mars as a god, named Ares, who carried with him the usual unfortunate associations. Though he was the son of Zeus, king of the gods, and his wife Hera, who was beautiful and wise, Ares was a wayward scoundrel and simply could not live up to his parent's accomplishments without getting into trouble. Perhaps the fact that Hera was also Zeus's sister may have been a factor—a troubled gene pool indeed.

The Romans had a similar interpretation of the god they renamed Mars. While they imported his virtues—or lack of them—from the Greeks, in Rome the red planet's penchant for violence was considered a big plus. Since the empire would conquer and rule by force, paying homage to Mars's psychotic tendencies made a certain amount of sense. The one thing the Roman Mars did have over his predecessors was a measure of intelligence—in many previous iterations, he was downright doltish. If the Romans made him a violent and warlike being, they also made him a smarter one. He went from being Steinbeck's Lennie to TV's Dexter in one sweeping cultural iteration.

Then Ptolemy, a Greek living in ancient Alexandria, came along in the second century CE and (surely with the best of intentions) created a model for the solar system that brought some apparent order to the chaos seen overhead, but also gummed up Western astronomy for well over a millennium. In his universe, Earth was at the center of all things, and the lights observed in the sky moved around it. These other bodies were affixed to spheres—unimaginably huge, crystalline, transparent ones—that rotated, imparting the motions observed. The obvious planets, those observable by the naked eye (Mercury, Venus, Mars, Jupiter, and Saturn), each had a crystal sphere of their own in logical order. The moon's was closest to

Earth, the sun fell between Venus and Mars, and the final sphere was that upon which the stars were mounted. This system, logical from a contemporary observer's viewpoint, was to remain fixed in place (just as Earth was, at dead center) for about 1,200 years. It was one of science's longest-held misconceptions, but at least it was an attempt at science. Despite this valiant effort at understanding the solar system, the broader culture that surrounded Ptolemy continued to see gods in the heavens.

Mars continued to be the scourge of the cosmic playground through medieval times, and not until 1543 did the red planet really become a *what* rather than a *whom*. Nicolaus Copernicus, a brilliant Polish astronomer, was finally able to build a mathematical model that supported a heliocentric system, that is, one with the sun at the center of the solar system (in fact, the sun was the center of the entire universe so far as he was concerned, but that inconvenience was cleared up much later). Then, almost a half century later, Johannes Kepler, a German scientist who worked out the laws of planetary motion, further developed the mathematical certainty of a sun-centered system.

About the same time, Galileo Galilei began looking the planets through his telescope, observing, among other things, the distinct phases of Venus. He was also able to see the planets as objects, not just as points of light, and their massiveness and motions under a heliocentric model made even more sense. He couched his opinions somewhat, though, as he had already endured enough torment at the hands of the Renaissance church to make him gun-shy. Speaking openly of his ideas turned out to be a recipe for papal investigation and, ultimately, a lifetime of house arrest. Sometimes it sucks to be too far ahead of your time.

It was not until the 1800s that Mars truly came into its own as a planet, apparently similar to Earth, in the minds of astronomers. While many were using telescopes at this time to observe the planets, notable among them was Giovanni Schiaparelli. During Mars's closest approach to Earth in 1877 (when the two planets' orbits bring them closest together, an event called "opposition"), he began a series of extensive observations that resulted in one of the first true maps of Mars. Schiaparelli invented a vast network of continents and seas for the red planet, further adding to its Earth-like mystique. Mars was continuing on its long journey to being a world, a place, that could be explored and—possibly—inhabited.

Among Schiaparelli's unintended "gifts" to Western science was the misinterpretation of his observation of linear features on Mars that he called *canali*. While in his native Italian this merely describes a channel (natural or not), to the English-speaking ear this sounds like "canal," and in a profound case of mistaken identity, a great controversy began.

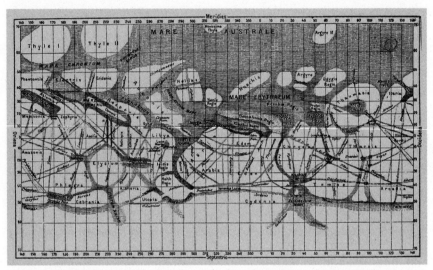

Fig. 3.1. *PLANETA MARTIS*: Schiaparelli's Mars map of 1877 demonstrated far more detail than he was able to see through the telescope. Many of the lines seen here did not exist, and were either figments of his imagination or the possible result of eyestrain. Regardless, these charts had a profound impact on popular thinking about Mars and strengthened the tradition of naming features in Latin. *Image from NASA.*

This misinterpretation was picked up and vastly amplified by an American amateur astronomer named Percival Lowell. Born of a wealthy Boston family with a fortune made in the textile business, Lowell spent years in Asia making a mark as a chronicler of Japanese culture before taking up astronomy seriously (he already had a degree in mathematics from Harvard). Once bitten by the celestial bug, however, he was unstoppable. He moved to the Arizona territory in 1894 and built an observatory in Flagstaff with the stated aim of observing Mars. Lowell commissioned one of the larger refracting telescopes of the time, an instrument twenty-four inches in diameter, to facilitate his Martian observations.

It's interesting to note that Schiaparelli and Lowell were somewhat contemporary to one another, with their respective peak activities being offset by about twenty years. When confronted with Lowell's interpretations of his work, initially Schiaparelli was somewhat miffed, but over time he softened and moved from denial to ambivalence, then even to a certain level of enthusiasm, about the idea of artificially created canals on Mars. Even then, people seemed to know the value of good PR. Lowell was the P. T. Barnum of Mars, and if you can't beat 'em . . .

Lowell took to his new life's purpose with relish. Over the next fifteen years, he spent countless nights in his cold observatory, observing and sketching Mars. Lowell also saw the elusive canals but perceived of them differently than did Schiaparelli. He charted these illusory features in much greater detail, numbering them and striving to discern some logical order to their design and placement. From these extensive and painstaking (and largely imaginary) observations, he began to develop his own theories about Mars and what—or *whom*—might exist there. Perhaps most profoundly, rather than limiting his ideas to scientific circles (where he met with opposition to his unique notions), he published a series of popular books between 1895 and 1908 that struck a chord with the general public.

These widely read books built a public case for intelligent life on Mars. Going far beyond mere observation and charting of the planet, Lowell imagined a vast civilization with a form of planetary government (after all, the rulers would need to have global reach to build such a vast network of canals). This technologically brilliant Martian empire was working against the clock to save their dying world as the water vanished. Mars was much older than Earth, he reasoned, and was thus farther along in its planetary evolution. It was cooling and was water-starved, and the Martians had built the vast network of canals, complete with pumping stations, locks, and other manipulations, to bring water from the poles of the planet to its parched temperate regions. It was an elaborate mental invention, and the romance of these ideas was to last for a half century in the public mind. Lowell's ideas about how Mars lost water were not really wrong, they were just off by about 3.5 billion years. The rest, engaging though it might be, was just so much mental popcorn.

While it is easy to dismiss Lowell's fanciful ideas today, reading his books (not a trivial undertaking) does impress one with the (admittedly misplaced) rigor of his thinking. If one starts off by accepting the idea that his Martians might actually exist, he builds a convincing case, using whatever scientific ideas were supportive to his arguments. While this is not a true implementation of the "scientific method," it was an attempt at building a logical view of another culture that held some allure.

Although few discuss it today, while Mars was wiling away the early twentieth century as a planet inhabited by parched, clever engineers, Venus was thought to be a sunny, humid jungle world. Little was known about the planet, and virtually nothing could be observed beneath its opaque cloud cover. Well into the 1950s, popular culture (aided by the likes of Edgar Rice Burroughs, who had his way with both Mars and Venus), Venus was the jungle world, with riotous plant growth and steaming swamps. The truth was even more shocking than it would be with Mars, as Venus was ultimately unveiled as a hellishly hot, sterile, acidic nightmare.

During the first half of the twentieth century, other scientists spent years observing Mars via the telescope, often striving to disprove Lowell's assertions. Using spectroscopes as their primary weapon, various observers were able to discern that there was far less water in the atmosphere than Lowell had theorized, and that the air was much thinner than appropriate for his Martian civilization. Lowell's romantic world was beginning to slip away.

This work was, however, restricted to the evidence of the telescopic eyepiece and other earthbound methodologies of observing Mars. Under optimal conditions, with a large, well-placed telescope, and even during its closest approach, the best image of the red planet is still subject to the vagaries of what astronomers call "seeing." Earth is covered in a blanket of air, turbulent and fickle, and air is dense enough to distort light. So even on the best of nights, the small, dim, red image of Mars swims in and out of focus, bending and flexing at the whims of the atmosphere. Profound or accurate observations of the Martian surface are difficult at best. The use of the spectroscope, a prismatic device that splits the light from a planet or star into its constituent gasses and elements, strove to overcome this limitation but was able to supply only a limited range of answers. Later, radio telescopes, unaffected by the optical properties of the atmosphere, were utilized to explore Mars but were also limited in what they could "see" on a planet and had other issues with Earth-based interference and background noise from space. In short, to really understand Mars, we would have to *go there*.

This idea of traversing the great blackness between Earth and Mars had appealed to many over the years. Fiction treated it in various ways: In the worlds of Edgar Rice Burroughs, writer of the John Carter books, all you needed to do was fall asleep in the back of an enchanted cave in the Wild West and * poof * you awoke on Mars. Later on, the brilliant Russian Konstantin Tsiolkovsky and Germans like

Hermann Oberth speculated on the use of rockets to carry people into space and perhaps even onto other planets. They were widely dismissed as cloudy-minded visionaries at best, crackpots at worst.

With the advent of modern rocketry in the twentieth century, these simple ideas became elaborate plans. No less a personage than Wernher von Braun, alleged Nazi and father of the Saturn V rocket that took America to the moon, labored on the problem. In his 1953 book *Das Marsprojekt* (*The Mars Project*), von Braun envisioned an armada of ten enormous spacecraft assembled in Earth orbit by reusable space shuttles. This fleet would embark sometime around 1965 for a three-year round-trip, including a yearlong stay on the surface of the planet.

The elaborate scheme was presented in a simplified form in *Collier's* magazine and generated a great deal of enthusiasm with the American public. This was further fueled by von Braun's appearances on episodes of Walt Disney's TV program *Disneyland*, which were of course set in Tomorrowland. The capstone episode for von Braun was titled "Mars and Beyond," and it seemed that we would be heading off to Mars any day now, preceded by a trip to the moon. It all seemed so simple, though vast and expensive. What was not known at the time was the harsh nature of the interplanetary environment, with astronaut-frying radiation that would accumulate during the long flight, as well as the fact that the Martian atmosphere was later discovered to be too thin to support his designs for winged gliders for landing the explorers. Still, it was a grand vision that stirred the imaginations of millions of Americans.

Then, in 1965, robots intervened, stealing the limelight away from the rugged human explorers of von Braun's imaginings. The truth about Mars would be far crueler—and bleaker—than anyone could have imagined.

CHAPTER 4

SETTING THE STAGE

As a youth I watched the American Apollo program's race to beat the Soviet Union to the moon play out, a contest the United States won handily. There had been a brief moment when it seemed the Russians might be the first to accomplish a manned lunar flyby, but the landing itself was never really in question. NASA's machinery, technology, guidance, and navigation techniques were generally superior.

After losing the space race to America, the Soviets claimed they were never really reaching for the moon after all, that they intended to pursue the peaceful exploitation of Earth orbit with a space station (a feat they accomplished in 1971 with their Salyut space station, two years before the United States matched the feat with Skylab). It was a significant achievement.

But there was another, quieter race between the global superpowers. The Soviet Union and the United States had been competing in space since 1957. In 1961, Russia attempted to fly Venera 1 past Venus, but the spacecraft lost contact with Earth after a week. Venera 2 suffered a similar fate, but the die had been cast in robotic exploration. In the same year, the Soviets did successfully orbit the first man in space, and in that case beat the pants off the United States. This rankled. NASA, at that point only three years old, was famously ordered by President Kennedy to land a man on the moon by the end of the decade. They were also more quietly given a mandate to explore the planets. A second, less public space race was born.

The United States reached Venus in 1962 with the Mariner 2 spacecraft. Along with a strong push to develop and fly robotic craft to reconnoiter the moon for Apollo, NASA ordered JPL to start pressing for Mars and Mercury as well as Venus. Of these, the Martian flights would be the most spectacular. This also threw down the gauntlet at the feet of the Soviets, who had Mars plans of their own. Their first Mars rover would beat Curiosity by forty years . . . sort of.

The flights of Mariner 4 past Mars in 1965, Mariners 6 and 7 in 1969 and the orbital mission of Mariner 9 in 1971 were all significant successes. In the same time period, the Russians hurled a number of probes toward Venus with the intention of landing on that hellish planet. From 1966 through 1970, no fewer than five landers failed in various stages of their missions (though Venera 4 did transmit data during its descent before failing). Finally, in 1970, Venera 7 landed on the surface, enduring the 900°F temperature and well over 1,000 psi pressure for twenty-three minutes before succumbing.

But Mars eluded the Soviets. In the same time frame, roughly 1960 through 1975, fifteen Soviet robots were sent to Mars. This shadow space race spawned a number of US attempts as well: six Mariner probes were sent to the same destination. But the results were spectacularly skewed: the US was four for six. The Soviets? Zero for fifteen. It was an amazing string of almost total failure for the Union of Soviet Socialist Republics (USSR).

Not so the United States. From its late start and shortly after that string of embarrassing second-place accomplishments, NASA quickly and quietly turned things around with the robotic cold warriors called the Mariners.

Of these, Mariner 4 merits recounting. It was launched just a few weeks after its twin, Mariner 3, failed due to a malfunction of its launch cover. Mariner 4 launched successfully and raced off to arrive at a point in space that would allow it to miss Mars by a few thousand miles, snapping images as it raced past the planet at interplanetary speeds. That was the hope anyway. With just one successful Venus mission under its belt, NASA could only plan and hope for the best.

Mariner 4 would succeed brilliantly but would not be kind to the romantics among us.

By this time, 1965, much was known about Mars. But these investigations had been conducted at the small end of large telescopes—some of which gathered optical light, some of which gathered radio signals. But all were firmly rooted on Earth and limited by that restriction. As previously stated, it was a tough way to observe, and in this era before computer-controlled adaptive optics and the like, it took a still night and a keen astronomer's eye to make much of what one could see. Also as previously mentioned, radio telescopes had the advantage of cutting through this atmospheric mishmash but still suffered from radio "noise," both from earthly sources and the great cosmos above. It too took skill to separate the wanted from the unwanted.

While much had been learned about Mars since the days of Galileo, it was nothing compared to what could be gleaned from even a fast flyby of the planet by a robot. But what had been learned by this time, the early 1960s, had already thrown a wet blanket on the followers of Lowell and others of his ilk. It was suspected that Mars had a much thinner atmosphere than predicted at the turn of the century— as little as one-tenth of Earth's, it was thought. There was also pitiful little water indicated by spectroscopic studies. Oxygen was almost nonexistent. What was *not* known was what might be seen on the surface once Mariner got close. There was a phenomenon that had been studied since the advent of larger optical telescopes in the nineteenth century called "the wave of darkening," which was a seasonal phenomenon that occurred roughly every two years (one Martian year). Each of these waves appeared to be related to polar ice changes and seemed to indicate that something was spreading from those regions to the equator on Mars. Was it plant life, as many had predicted over the decades? Was it a series of dust storms caused by seasonal changes? Or was it perhaps merely an illusion, as others thought? This query would have to wait for the robots to do their work. One thing did seem certain: the canals, so often observed in the nineteenth century, were far less apparent in the twentieth. It was not good news for Lowell's believers.

It is worth noting that the optical coating of lenses, that is, the depositing of a thin layer of special elements on lenses to cut down on reflection, did not come into wide use until the twentieth century. This may be what first killed the canals of Lowell and Schiaparelli—it has been postulated that the nineteenth-century astronomers may have actually been mapping the reflections off the eyepiece of the capillaries in their retinas. The antireflective coatings of later years may account for the reduction in observations of canal-like features—we will never know for sure. But Lowell would surely be ashamed if he realized that his vast empire of brilliant Martians was actually just meandering blood channels in his eyeballs.

A personal note—I was nine years old when Mariner 4 flew past Mars, and I followed the mission carefully (as carefully as a nine-year-old can follow anything). After all, JPL was only about three miles from my house. In fact, I later discovered that my childhood home was within the blast radius of Soviet nuclear warheads aimed at both JPL and Caltech—that's kinship. In any case, I felt a sense of connection with the robotic programs being conducted from there. Of course, as a youth steeped in the science-fiction movies and books of the day, I was wedded to

Lowell's romantic fantasies as much as anyone had ever been. My father had taken me on a trip to Lowell's observatory in Flagstaff, Arizona, and there I had seen the gigantic (to a nine-year-old) telescope Lowell had used to create his elaborate illustrations of Mars. My primary memory besides the amazement I felt when I saw this wonderful, polished, brass and steel star-cannon was my surprise that the massive wooden dome that housed it rotated on a series of automotive wheels and tires. It seemed an ill fit for the grandeur of the enterprise.

At the time, Lowell Observatory was in the odd spot of wondering what Mariner might find on Mars, and how it might compare to the vision of its founder. Astronomy had long since moved on from Lowell's Mars, but some, like the readers of the illustrious Ray Bradbury and his 1940s fictions about Mars, were steadfast— it wasn't over until it was over; the fat lady had not sung; and Mariner had not yet flown past the red planet. There was still a chance, no matter how small. But the red sand was flowing through the hourglass, and time was running short.

On July 14 and 15 of 1965, Mariner 4 made its kamikaze dash past Mars. The magnetometers searched for a magnetic field (and found little), the radiation detectors recorded their findings, and the TV camera snapped 22 and 1/2 grainy, low-resolution images of the planet as it hurtled by. Data were recorded to an onboard tape recorder for later playback, since transmission rates this early in the digital-imaging game were dreadfully slow.

Back at JPL in Pasadena, the numeric data were coming down, but it would take much longer to photographically print the ghostly black-and-white images. Impatient scientists took some of the data printouts, which were, after all, numeric indications of gray values, and with a set of colored crayons made something like a paint-by-the-numbers kit and created the first close-up images of Mars.

A small historical aside: when the guys in the white shirts and skinny ties wanted to color-in the printout, they had nothing but gray pencils at the lab, so they ran down to a local Pasadena art store. The snotty clerks there (they were still that way when I went to art school in the 1980s) told them that they did not carry actual *crayons*—those were for little kids. They did, however, carry greasy pastel paint sticks. The engineers had their choice of yellow-orange flesh tones, or blue and green. They opted for flesh tones—otherwise, our first color images of Mars might have looked more like a blob of algae. That resulting hasty drawing, in all its fleshy grandeur, still hangs on a wall at JPL today.

Then the actual pictures came out from the darkrooms. The aforementioned scientists gathered 'round, passing the prints from person to person while they were still dripping wet. They were dim, fuzzy, and monochrome . . . but to them, beautiful. Here was proof at last that Mars was not home to canals, forests, or oceans, but that it was indeed a dry, apparently lifeless desert. The images showed craters not unlike those seen on the moon, and little else. The romantics had their images of a fecund, inhabited Mars shattered in an afternoon. The cities, the oceans, the planet-girdling canals built to save a dying culture vaporized. It was a bad day to be a romantic, and I remember it with a bitter taste.

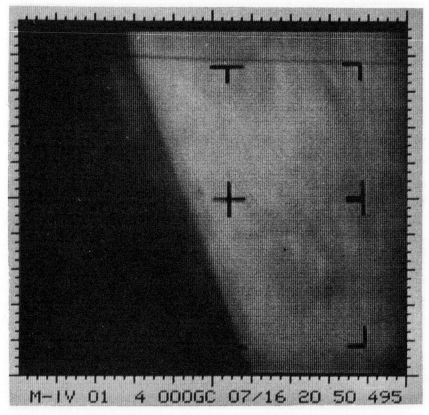

Fig. 4.1. FIRST FROM MARS: This is the first image transmitted by Mariner 4 in its 1965 dash past Mars. The 22 1/2 images it sent back were low-resolution, indistinct . . . and enough to convince the scientists that Mars was as dead as the moon. *Image from NASA/JPL-Caltech.*

Four more Mariners would be shot into space for a rendezvous with the God of War. Mariners 6 and 7 were also flybys, reproducing the mission of Mariner 4 with greater precision and passing far closer to the planet (for those of you who are counting, Mariner 5 went to Venus). But it was the final Mariner effort that really stunned the planetary-science world.

In a replay of the Mariner 3 failure, Mariner 8 aborted its Mars mission when the rocket booster failed and exploded. Mariner 9, which had been programmed with slightly different objectives than its twin, would now have to do the work for both, mapping the red planet from its equator to the poles.

Mariner 9 reached orbit without undue drama in November 1971 and sped off to Mars. But this was a Mariner with a difference: it was far larger and heavier, and it carried a quantity of rocket fuel that would allow it to brake into Martian orbit. On November 14, 1971, it did just that and entered orbit around Mars . . . right in time for a planet-girdling dust storm.

When the storm abated a month later, JPL turned the cameras back on and started imaging the planet from close up. First, three massive circles appeared above the settling dust—and the planetary scientists realized that they were looking at the largest volcanoes ever seen, on any planet. Then, as the clouds continued to clear, more and more features swam into view. Within a few months, the machine had mapped a goodly portion of the planet, and what the scientists saw stunned them.

The dead, arid Mars suggested by previous flights was now visible in stunning resolution (for the time), and a whole new story began to form. There were what looked like ancient river deltas, alluvial fans, canyons, fissures, and other features whose origin clearly involved sculpting by water, and lots of it. Lowell's romantic (and apparently once-wet) Mars recovered by one tiny increment, and we now saw it as a complex and still-active world.

Then the Viking program, a mission as audacious as it was evolutionary, took final form. In the making since the early 1960s, these twin massive probes—each with an orbiter and lander—would head off to a rendezvous with Mars in 1976. The orbiters would continue the work of the Mariners: map the surface, search for a magnetic field, and in general explore the regions above the planet. But the landers would be something brand-new and amazing.

It should be noted here that the United States was not technically the first nation to land on Mars. The Soviet Union had enjoyed a string of early successes (after a fashion) with the exploration of Venus. That planet—both closer and easier to navigate to from Earth—was to become to domain of Mother Russia for quite some time. The USSR had an early advantage of larger rockets than the United States, and their probes to other worlds reflected this. Huge, bulbous, and bizarre looking, these spacecraft were heavy and unwieldy, but the Soviets had lifting power to spare, so it was a nonissue. As it turns out, the bulkiness of the Russian spacecraft had a simple origin: their electronics were not up to the task of operating in a hard vacuum (as the US spacecraft had done since the beginning), so the Russians *pressurized them*. In short, they created the equivalent of small, pressurized space capsules to house their electronics and protect them against the harshness of space. As a result, the machines looked as much like a disease as they did like planetary probes, but they worked—sometimes.

So the Soviet Union had flown a chain of tough spacecraft to Venus with some success. But when they sent missions off to Mars (which they did with staggering regularity) the game changed. Mars is *hard*, as more than one JPL employee has stated with chagrin, and has remained beyond the scope of Soviet (and later Russian Federation) technology for decades. But when one of those doomed Russian spacecraft, Mars 3, joined Mariner 9 in Martian orbit, five years before the Vikings arrived at the red planet, a Soviet lander detached from the orbiter and managed to set down on Mars. Once on the surface, the machine began to take a picture to celebrate the triumph of the Worker's Paradise over the decadent West—and then it died, having transmitted just seventy lines of video gibberish. It survived there for a scant fifteen seconds. But the Soviet Union did not give up easily (ask a German soldier from World War II) and continued to send a string of spacecraft to Mars for decades. None have made it to date.

Of interest to the American Mars rover planners, the first rover on Mars was not actually Pathfinder's Sojourner, as is so often stated. The Mars 3 lander had a tiny rover on board, which was truly revolutionary for its time. It looked like a maimed Hoover vacuum cleaner without the handle, was attached to the lander with a roughly forty-five-foot cable, and was designed to gimp along on two skids that would lift, move forward, and nudge the rover along with them. It was actually an ingenious system and, while limited in scientific scope, was well ahead of its time. Unfortunately, it died along with the Mars 3 lander.

The United States, meanwhile, had been spending copious amounts of money designing and building the Viking spacecraft, especially the landers. Not content with making better maps from orbit, and not satisfied with landing there merely to perform basic surface analysis, Viking's designers upped the ante substantially— they wanted to look for life. The results of this life-science orientation would set the stage for every Mars surface mission to come, and would ultimately steer mission decisions for Curiosity.

Fig. 4.2. BRILLIANT BUT FLAWED: The USSR's Mars 3 mission was comprised of an orbiter, a lander, and a tiny rover sent to Mars in 1971. The lander is seen here. It could have been a spectacular all-in-one mission, but design flaws and bad luck resulted in useless pictures of a dust storm–covered planet and a dead lander. *Image from NASA.*

The Viking program would enjoy a budget that later JPL program managers could only fantasize about. While the mission was flown six years after the semi-official end of the space race, it was proposed, designed, and built at the height of that contest for technological domination. The $1 billion budget would equate to something like $7 billion today . . . a truly vast sum by current standards and one we are unlikely to see again for a single planetary program (perhaps if they found a pyramid on Mars . . .). But during the Apollo years, it was just a trickle from the roughly 5 percent of the US budget spent on the overall space program.

The landers contained mini life-science labs that were miracles of miniaturization. This was more of a feat than it might at first seem. In that day and age, such instrumentation usually filled a large room. The Viking lander's laboratories were about the size of a dishwasher. Inside were four experiments: three designed to look for microbial life and the fourth to perform basic atmospheric and soil analysis. But it was the biology experiments that ultimately caused such consternation.

Scientists come in many stripes, but among the prominent ones there is a tendency to have an equally prominent set of beliefs. Of course, the very basis of scientific inquiry requires a mind open to the results of experimentation and an acceptance of the results. But this does not preclude a strong attachment to hypotheses, and in this the Viking crew were divided roughly into two camps. One was the "I doubt we will find anything, but let's load some soil into a container, expose it to some carbon-14, pump out the air, burn the dirt, and look at what comes out of it" camp; the idea was that living things would metabolize the ^{14}C and the instrument could then measure it.

The other camp also started with a soil sample but then added a nutrient "broth" to it, thought to be something Martian microbes would eat as readily as earthly ones would. This was the "grab the dirt, feed it, and see what shows up" crowd. Once wetted, the sample would be monitored to see if it gave off methane or any other by-products of metabolization.

There were two other soil experiments: another form of feeding the dirt (and anything within) to measure resulting gasses, and yet another that would look at untreated soil samples that were heated. But it was the first two that caused the controversy—the two men in charge simply could not agree on much. The "I doubt we will find anything" group was led by Dr. Norman Horowitz of Caltech. The "Let's feed it and see" camp followed Dr. Gilbert Levin. The opposing results of their respective experiments would fuel a multidecade controversy and affect the design of Mars landers and future experiment packages heading off to Mars for decades.

The landers set down on opposite sides of the planet in 1976, Viking 1 on July 20 and Viking 2 on September 9. Each soon deployed an arm and scoop to gather soil samples, which they then deposited in small, onboard receivers. Each of the experiments mentioned got a bit of sample material and went about their tasks. In general, they created differing conditions within the samples and then looked at the results to see if there were any positive indications of microbial activity. When the first samples were ingested and processed, three of the instruments indicated readings consistent with sterile soil—no life found. The forth, however, demonstrated a rapid spiking of what could have been metabolic activities, then declined just as quickly. Researchers were excited but then puzzled by the odd readings; they should not have declined so precipitously. In the end, the general consensus was that it had been a mere chemical reaction in the soil with no microbes in sight. A small faction, however, continued to interpret the reading as indicative of biology at work. The debate has quietly raged ever since.

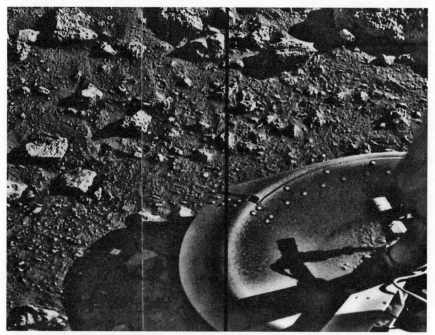

Fig. 4.3. SUCCESS: The first image successfully transmitted from the surface of Mars: in 1976, the Viking 1 lander sent this image back to a gleeful JPL. It shows the footpad firmly planted on Mars, assuring nervous flight controllers that the lander was level and safe. *Image from NASA/JPL-Caltech.*

In the meantime, the thirty-foot-wide Viking orbiters wheeled overhead, snapping images of the entirety of the planet. Between them, the two orbiters returned tens of thousands of images of the Martian landscape, mapping its geological complexity in exquisite (for the time) detail. Mariner 9's discovery of massive erosional forces that had at one time been at work below were confirmed and scrutinized in detail. Where all that water had gone was still a mystery, but hundreds of gullies, gorges, buttes, and deltas had been formed by something vast, violent, and wet in the ancient past. Future spacecraft would unravel the mysteries catalogued by the long-serving Vikings.

One by one, the orbiters and landers shut down, with Viking 1 being the last to succumb in November 1982. It was not a victim of the harsh environment but of an erroneous command that caused the radio dish to rotate away from Earth.

That day in November would be the last time a signal would be heard from the surface of Mars until 1997.

CHAPTER 5

THE SCARECROW

We're taking a break from the history lesson that leads us to MSL and the Curiosity rover. After that first memorable expedition to Death Valley, I was invited for a second visit, but this one would place me within a few hundred yards of my car at all times. I could hardly complain. The activity was some sand-dune testing of a variant of the Curiosity rover specifically designed for driving tests on Earth. They called it the Scarecrow.

The Scarecrow was a smaller, lighter version of Curiosity. The resemblance to the real thing was vague, but functionally it was identical, at least so far as driving on Mars was concerned. Its lithe design, just a framework, some motors, and six wheels, was designed to mimic the way Curiosity would handle driving in the 0.38 of Earth gravity that it would face on Mars; Scarecrow weighed only 38 percent of what the real rover did and should interact with the sandy terrain in a way similar to what Curiosity would experience on Mars.

Oh, and the computer is off-board, connected by a cable. That's why they call it the Scarecrow—it has no brain. Get it? I guess that would make John Grotzinger the wizard, but he might not see it that way.

I should mention here that although I have referred to Grotzinger as "the Big Kahuna" and as a wizard, John himself would assuredly defer any kudos to the team of scientists, engineers, technicians, and support people that make up the MSL mission. JPL'ers are team driven, and few will take single-source credit for an invention or an accomplishment. So whether you are talking to Grotzinger about the mission science, to Manning about the engineering, or to Vasavada about planning and execution, they are likely to remind you that it is a team effort and that they just happen to be in charge of all or part of that particular team. And it's true. But we civilians tend to seek out a person in charge of some part of the mission, a personality with whom we can identify, a leader with which we can share a moment of glory. Blame it on centuries of training by organized religion and, more recently, supermarket tabloids. But for the individuals whom I spoke of and others profiled in this book, I'm sure that they would have me remind you one more time: it's a team effort.

Back to our tale.

It was still months before the launch, and this was a chance to see the hardware at work. I drove out with John Beck-Hoffman, the JPL videographer who had joined us for the Death Valley adventure, and we discussed video production and space exploration during the multihour drive. It was fun to get another perspective. "I've worked at JPL for over twenty years," he told me, "and still learn new stuff all the time." We discussed his now-famous "7 Minutes of Terror" video, which he wrote, shot, edited, and even scored. The man is irritatingly talented. He has since left the lab, surely a loss to their media efforts. As Grotzinger once commented, "That video is the gift that just keeps on giving. I cannot give [a] public presentation without showing that thing. When I look at that, it brings a tear to my eye and for the audience it's just another hooting and hollering moment." NASA can use all the hootin' and hollerin' moments it can get.

Fig. 5.1. IF I ONLY HAD A BRAIN: The Scarecrow, a stripped-down version of Curiosity for use in mass-equivalent (i.e., simulated Mars gravity) testing, is to the right. Mike Malin, creator of Curiosity's camera systems, is being interviewed to the left. *Image from NASA/JPL-Caltech.*

Everyone loves an audience. Beck-Hoffman got one, many millions strong. It was a spectacular piece of outreach.

When we arrived at the test area near the west entrance to Death Valley, the Scarecrow was already up and running. There was a tarp-covered spot of shade where a few young engineers operated the rig from a laptop. Grotzinger was here, then over there, then back again, making sure that the rover would work in the way it was expected to.

A gentleman named Mike Malin was there too, working off the tailgate of a truck, testing some bits of camera equipment. Middle-aged, bearded, and not slim, he could pass on a quick glance as my brother. An area where nobody would confuse us would be in intellect. I get by fine, but Malin has a MacArthur Foundation award under his belt and owns a mind of that caliber. You find a lot of that around Caltech and JPL. He speaks expressively but does not dwell on himself—it's hard to tell if it is modesty or simply that the subject at hand is more important. I suspect, in his mind, it's some of both.

Watching him tinker with the cameras that day, it was pretty clear that he is always working on something, often a new invention to enhance planetary photography. Talking to him makes it clear that (a) he is a genius and (b) he does not suffer fools, or the press, gladly. Fortunately, Grotzinger was kind enough to make an introduction and that won me an enjoyable conversation. "Rod's one of the good guys . . ." he said, and I heartily agreed. I was on the right side of the angels that day.

I asked Mike what he was doing out there in the windblown sand and scorching sun. "We've been out here all week doing a field test of the mobility system on the rover to determine how well it will handle navigating on sand. We're just finishing up a couple of tests looking at intermediate slopes and intermediate compaction of the sand." As you might imagine, sand comes in all sorts of conditions on Mars (steep dunes, shallow dunes, fields, pits, mounds, etc.), as well as in varying levels of compaction (soft, hard, medium, etc.). And on Mars, with only 38 percent of Earth's gravity, the slopes are at different grades and the sand can behave differently. Testing is paramount.

"We got through the extremes yesterday. We did some testing in a flat area with the sand all churned up, and then went to steeper grades to see how well it could handle steep slopes of sand. Today's test is in between those extremes to get some transition information, and then the last thing we will do today is to try to drive it up a dune over here and see how far it goes before it stalls." The dune is not terribly steep for a human walking in one gravity, but for a machine like Curiosity, in Martian gravity, it could be a different story.

Clearly this is more than just work—Malin is here for more than the technology: "I'm out here for a couple of reasons. First, this is fun for me, I am a geologist, so this is my natural environment and habitat. Most of the time I'm sitting in an office banging away at a computer, so to sit outside and actually feel the wind on my face and watch the sand blowing is very soothing. Generally I'm here to advise the engineers on how and where to test the rover, but this is really vacation for me."

I then see Malin's associate fiddling with a couple of off-the-shelf digital SLR cameras connected via a cable to a computer rig, and query Malin about them. "This is a Mastcam simulator." Mastcam is the imaging system atop the camera mast on Curiosity that Malin built. "It's a consumer-level version of what we flew to Mars, so it's very inexpensive to build. It has two cameras, each with 35 mm and 100 mm lenses. The whole assembly is secured to a mount which we got from a telescope and which is controlled by this Mac mini. We write sequences that look like the ones we send to the cameras on Mars, and we interpret those using this consumer-based camera by having it do the same process."

His assistant had rigged the Mac mini in a drink cooler with ice to keep it from overheating. Smart—I could have used one of those on my first Death Valley excursion, for my head. Mike continued: "As with most engineering you want to take it outside where you will be working in the real environment to test it. Sometimes it might not work all that well. We intend to test it in other environments later this summer and into the fall. We are using it to simulate what we will be getting from Mars as a training tool, then when we get to Mars we will use it to take similar pictures so that we can compare what we see on Mars to what we see on Earth, and we then learn from those comparisons."

I thought the rig was pretty ingenious. When we hear of space missions that cost $2.5 billion, we tend to think of all the work as being conducted the same way that Apollo was—with expensive, top-drawer, custom technology, all work done in clean rooms. In reality, the only way to keep a mission like this as inexpensive as $2.5 billion is for people like Malin, who has probably contributed double the time than he has charged for, to work with off-the-shelf components doing clever tests like this one.

After this chat, and a closer look at his impressive rig, I wandered over to where the Scarecrow was doing its thing. The machine was slowly working its way up a sand dune, moving at a glacial crawl, which is Curiosity's usual speed of transport. It's slower than you think: quicker than the minute hand of a clock but possibly more sluggish than the second hand.

Along its route were set up a couple dozen still cameras on tripods. The Scarecrow had four large, white, foam balls spaced across its top, one on each corner. They were being used in the same way that movie visual-effects practitioners use Ping-Pong balls on actors in front of green-screen setups to track the motion of the subject, in this case the Scarecrow. This documentation aided in motion studies and tracking any slip in the sand. Slipping is more critical than it might at first seem, for reasons we will learn soon.

Based on experience gained from the Mars Exploration Rovers (MERs), especially Spirit, sand dunes and sand traps are a big concern on Mars. Spirit got bogged down a couple of times, despite its own exhaustive testing. This was exacerbated by the malfunctioning of two of its six wheels. The last sand trap was fatal, and Spirit died in a grainy bog. It was a lesson well learned, so extensive studies were being conducted across a range of sand-related questions: How steep a slope can you assault before the machine begins to slip? Under what conditions will it become bogged down? What is the optimum speed for it to travel on different surfaces? Are there specific kinds of sand environments that it should avoid altogether?

While the engineers were working out their troubles, I had just discovered a few of my own. My video camera was acting up, probably due to its being made of black plastic and the hot sun beating down on us; it had become a twelve-pound heat-sink and was quite hot to the touch. I covered it, and once it cooled it ran fine, but then I discovered another, more-insidious problem. There had been a light breeze all day, which felt wonderful in the hot weather. But that breeze, upon closer inspection, also carried a fine, almost-invisible cloud of sandy dust that, weirdly, stopped at about shoulder level. I hadn't really noticed it. But when I went to twist the camera's focus ring in midafternoon, I heard—and felt—a *gruuuunch!* sound. Sand had gotten into the lens mechanism. Fine, sharp grains of sand and optics don't mix well, and the proceeds from this assignment would be more than devoured by the trip to the repair depot the camera would be making.

Oh well, keep shooting. It's all for the space program. And to be honest, I was having a lot of fun—it was a space geek's delight.

Toward the end of the afternoon, I got together with Grotzinger for an interview. By this time I had not only shaded my camera but also had wrapped it protectively in a plaid shirt (the shirt was wretchedly ugly; I was glad to put it to one more use before I tossed it). This engendered a couple of odd looks from the press corps, but I figured they would eat crow when they got home and later found their own optics going *gruuuunch!* So much for camaraderie in journalism.

Too soon, the press interview drew to a close and we were shooed-off to allow the JPL folks to get some work done. Beck-Hoffman and I drove back to LA. He would soon be off to work for National Geographic TV and other outlets, and I had a book to write.

Scarecrow continued to work for the rest of the week. Alas, it never found a brain, but did lay a solid foundation for Curiosity's upcoming travels.

CHAPTER 6

LAYING THE FOUNDATION: MARS PATHFINDER

The first rover to Mars arrived in a somewhat-undignified fashion. After a mad dash from Earth, it smashed into the thin atmosphere, cut a scorching swath across the sky, popped a parachute, fired some braking rockets, and then bounced. Quite high, and again and again, because it was designed to do so. But even if it was inelegant, it worked.

The landing of Pathfinder was exciting and arrested the attention of the public. This plucky little machine, developed quietly and on a limited budget, became the darling of the Internet. JPL's servers crashed on the first day from the millions of hits. This was the first machine on Mars since 1976's Viking landers, and it thrilled the world for the next three months.

But, prior to launch, Pathfinder had been a touch-and-go mission. Many inside the space agency had opined over the years that rovers were too complicated, were too risky, and would be too expensive and too hard to land. But by the early 1990s, technology had progressed sufficiently so far downstream since the Vikings landed on Mars that it was clearly time to reconsider. A small team in JPL came up with a new approach, that would ultimately become the first wheeled vehicle on Mars, a tiny precursor to Curiosity. And the public loved it.

It was a right-sized mission for the time. NASA's new administrator, Daniel Goldin, had decided that the agency needed to stop building large, heavy, and multipurpose spacecraft like Galileo (Jupiter) and Cassini (Saturn). The school-bus-sized behemoths were expensive to produce, launch, and operate. And a single failure, as almost happened when Galileo's antenna failed to deploy fully, could mean a billion-dollar loss. In the financial climate of the day, the continuation of these flagship missions was not high on the list of preferred projects.

Enter the Discovery Program. Goldin's mantra was to make space projects "faster, better, and cheaper." It was a tall order, for in space exploration (and technology in general), those three things do not often play well together. If asked, many JPL engineers would have told him, "Sure, pick any two." But in this era of lean budgets (not unlike today), if you wanted to fly a spacecraft, especially if it was "interesting" (read "risky"), then the Discovery Program was the way to go. The idea was to design and build Pathfinder as a mission costing less than $150 million; all Discovery Program missions would have constrained budgets.

A man named Tony Spear was Pathfinder's project manager, and Matt Golombek was the principal investigator. Spear assembled a small team of young and energetic engineers and scientists within the lab to design and build the spacecraft in a skunk-works-type environment. The overall budget, including flight and operations, was to be capped at under $300 million. When this is compared to the design-and-build price tag of half that, it shows you just how expensive it is to launch and operate these missions—in this case, it doubled the cost.

Pathfinder would be comprised of two components: a lander and a rover. The lander, looking like a small, metal pyramid, would set down on Mars. Then the sides would drop, and a tiny rover, no larger than a microwave oven, would drive off and explore the locale. They named it Sojourner. It would only range about thirty feet from the base station and had to use the lander's radio to talk to Earth, but this was a groundbreaking mission, and getting *anything* onto the surface of Mars would be a major accomplishment after two decades of no Mars missions. The team took to the task with passion.

Many years later, I sat with Rob Manning, who had been a pivotal character in the Pathfinder saga. When we talk Mars, there is a gleam in his eye and he gets so excited about space exploration that he can barely get the words out—but he does exclaim that this or that bit of technology "is just *fantastic!*" quite a lot. I sympathize—I don't possess half the knowledge he does, but I too get impassioned about the subject, and it's something I've been taught by long public exposure to limit in myself—some people find it slightly *frightening*. But with Rob, it's a charming combination, as witnessed by the number of JPL people who stopped by our table as we dined. We had visited a favorite eatery not far from the lab, and at least five groups of people wandered over to check on something project oriented, ask him a question, or just say hello. Each was greeted with exuberance. It's as if he is permanently just out of Sunday school, in an exhilarating way.

Fig. 6.1. "JUST FANTASTIC!": Rob Manning was the chief engineer of MSL until he moved to his current assignment as chief engineer of a new, high-speed EDL (entry, descent, and landing) research program for future Mars missions. Rob is the kind of guy who gets other people excited about Mars. He does frequent speaking events, and his enthusiasm is incredibly contagious. He led the design team for Pathfinder and has worked on every Mars rover mission to date. Of Pathfinder's bouncing arrival, he remembers that it had been "tough to pass the laugh test." *Image from NASA/JPL-Caltech.*

I took him back to the beginning phases of Pathfinder in the early 1990s to learn more about the machine that would ultimately provide many core technologies for Curiosity as well as the Mars Exploration Rovers, Spirit and Opportunity. As it turned out, it was the fastest and by far the simplest mission he ever worked on . . . partly because NASA's expectations were low.

"The great thing about Mars Pathfinder is that we had a single page of requirements, the government just said 'land on Mars during the 1996 launch opportunity, deliver the rover and send back some pictures. Good luck! Here's some money to do some science if you have the time.' This kind of mission design had really not been heard of since the 1960s. It was delightfully freeing." The memory brings a smile to his face.

"Basically the mission model was to prove that NASA can do things cheaply, efficiently, and effectively. Also to demonstrate that there are efficiencies that we can achieve using the faster, better, cheaper approach which had not been really invented yet. In fact, nobody had ever really agreed what that meant. . . ." He chuckles.

The first decision was to have a team working together, all in one spot, at JPL. "We thought that was very important. We couldn't completely achieve that because there were space [as in office space] problems at the lab at the time. But we got the majority of the core team members all in one spot." This allowed for close coordination, rapid communication, and a delightful lack of paperwork. "There really was very little documentation," he says with a smirk, "there just wasn't time!"

Perhaps the largest challenge, as it would be with Curiosity, was how to deliver the spacecraft to Mars. Viking, with its vast funding, had carried along large tanks of rocket fuel not just for the lander but also to allow it to enter orbit around Mars. When it arrived in the neighborhood, it fired powerful braking thrusters to slow it enough to be captured by Mars's gravity. The twins orbited the planet for weeks, while at JPL scientists fretted over how nasty the landing zones looked in the improved pictures. But they picked a spot remarkably close to the original one, and landed, falling out of the sky at mere orbital speeds.

Pathfinder was an entirely different animal. It was smaller and far lighter, true, but with their tiny budget (about one-twentieth Viking's when inflation is accounted for), they would be unable to pause in Martian orbit—they simply did not have enough rocket power to stop there. So they targeted the probe to fly directly into Mars's atmosphere. It was a lot like shooting an enormous hunting rifle at where Mars *should* be in eight months—with no allowance for error.

This meant that Pathfinder would slam into the Martian air at interplanetary-transit speeds of about 12,000 mph, way faster than Mars orbital speeds. It also meant that the heat shield needed to work harder, and they had more energy to get rid of on the way down. Otherwise—splat, no more Pathfinder.

To accomplish this, they chose . . . beach balls.

Rob explains the original idea: "When we hit the ground, you don't want to store energy and rebound. So the original idea was to have vented airbags like the ones in cars that would deflate right away, so all this energy is dissipated and is more like a sandbag." In short, once the lander had slowed to manageable speeds, they would inflate airbags—looking very much like a collection of three-foot beach balls—to absorb the impact of landing, and they would immediately exhaust the air out the sides upon impact. He continued: "But then we realized we couldn't get the vents to work quickly enough . . . so we all agreed that we would be bouncing instead." They would scrub off the remaining landing energy—its speed—by bouncing until the spacecraft rolled to a stop. Then the airbags would be deflated.

While the project generally ran somewhat under the radar, they did have to submit to the usual NASA design review. This is where the best minds on the project develop their sales pitch, then meet with a group of engineers, scientists, and others from the field to sell their big idea. On the panel for Pathfinder were some Viking people (that was good), an assortment of current NASA folks, and one somewhat-curmudgeonly Cadwell Johnson (maybe not so good).

Johnson was an interesting choice. While he was nearly a god in the aerospace community, he did not even have a college degree. He had earned his stripes before the modern era of NASA where it is generally recognized that a master's degree is a point of entry, and PhDs are common. But when Johnson came into the space program, at the dawn of NASA in the 1950s, they just needed his pencil. He was one of the best and brightest draftsmen that Max Faget, designer of America's manned spacecraft from Mercury through the space shuttle, had ever met. Johnson was hired on the spot and became a legend of the space race. And now he was staring down a somewhat-nervous JPL team.

"So here we were are, having this review, and Cadwell is listening impatiently. Richard Cook, the flight-operations engineer, was talking and about mission design and defending the airbag architecture." Manning smiles at a memory that brought little joy at the time. "Cadwell says, 'Listen buster, don't tell me how to land on another planet! This is a stupid idea that's never gonna get off the ground!' He was very negative. And the Viking people were rolling their eyes. They're rolling their eyes and they are saying 'You're gonna do *what*?'" He laughs his infectious laugh. "At that point, we had told them how high we were bouncing—we're talking fifty to seventy-five feet above the surface of Mars. We were testing it to be able to bounce *one hundred* feet. They just thought we were nuts."

As we all know, it turned out to be a sound design, but the review was a confidence rattler. Nonetheless, it was eventually approved and they moved forward. The lander and rover were designed and built at JPL by a small and dedicated crew of engineers and technicians. The parachute was a headache, though—all parachutes are headaches when used to land on Mars because fabric does not behave like metal or plastics. "It kind of does its own thing," as Manning says. But the real challenge remained the airbags. Now that they had sold the landing system, they had to make it work.

"NASA had said to us to keep it simple. Don't overdo it, don't overspend. . . ." In short, stay within the boundaries of the Discovery Program. "Keep it simple— that was our mantra." Actually, there was another, even more urgent imperative— don't screw up. "The other mantra was to test, test, test," Manning recalls. "We did not skimp on testing. Sometimes the test articles were low budget, but we didn't skimp on test itself. We had to convince them, and ourselves, that we knew how to get [this landing system] to work."

Testing would be a key ingredient in the success of all the rovers, from Sojourner to Curiosity. This remains the only way to run a successful program.

There is a lot of jargon in aerospace engineering, and one part of it is around the readiness and flightworthiness of a given design or system. It's called the DRL for Design Readiness Level. The design is assigned a number to indicate its reliability ranging from 1 to 9. "I would say that airbags were probably at a 3," Manning continues. "They're supposed to be a 6. We did a lot of things for the first time that were later adopted by other projects, technologically speaking. But we weren't stopped from being innovative, because it was considered part of our objectives."

They tested and they tested. The bags ripped and tore. They inflated late and early. The engineers rolled them down inclines at JPL and dropped them on the desert floor from a helicopter. It was all somewhat *ad hoc*, though thoroughly professional. Just . . . small. But over time they wrestled the airbag system into compliance.

The rover itself was an innovative design. The small box was topped with a solar panel that provided just enough current to allow the machine to work in various lighting conditions. It was mounted flat atop the little machine, however, and one good coating of the ubiquitous Martian dust would reduce its efficiency something fierce. But it was the only way it would fit. Sojourner sported six wheels, each about four inches in diameter with little aluminum spikes jutting out of them for traction. The wheels were mounted to an ingenious suspension system called "rocker-bogie," a jointed set of levers on each side that allowed maximum ground contact when driving over obstacles. This suspension system worked so well that it has been used on all Mars rovers since.

Of course, the lander and rover themselves also had to be tested. While the lander was fairly straightforward, with its landing bags and unfolding sides, the rover needed to be *driven*. And on something approximating Mars. And . . . on the cheap.

"In the center of the JPL campus is a building called the SFOF—the Space Flight Operations Facility. Upstairs is one giant room that extends the length of the whole building, from corner to corner. That giant room was committed to Mars Pathfinder, so I had sandboxes installed in there." There is a playful, possibly mischievous glint in Manning's eye as he relates the story. "We had to do a lot of arm-twisting to let me bring in thousands of pounds of sand for the test area. In the end, I bought playground sand from Monrovia [a sleepy suburb fifteen minutes east of Pasadena, where things tend to be less expensive]. We washed it first so it wouldn't be too dusty, but we still had to seal off the room to keep the sand inside. So we had our own little sandbox up there as well as the electronics, the lander, and airbags. All in one place where the team could work together. It was great fun." I can just see the administrators gleefully enjoying the idea of a few tons of sand next to their überexpensive operations center. . . . Not.

And in one more trip to Monrovia he visited a Home Depot store and bought a leaf blower. "I had to use that to inflate the airbags for the early tests." He laughs.

The lander and rover went through tough testing at the lab—mostly successfully. This included thermal testing: the little machines were roasted, then frozen repeatedly in JPL's thermal chamber. They were shaken to make sure that they could survive the savage rocket ride into orbit. Then they were subjected to a thorough cleaning to avoid contaminating Mars with earthly bacteria. It was a grueling process, and all done on a tight budget.

The entire assembly left Earth on December 4, 1996, aboard a Delta II rocket—the only US rocket powerful enough to do the job on their budget. Pathfinder transited the cold dark between Earth and Mars for seven months, then slammed into the Martian atmosphere on July 4, 1997. The heat shield, the first US design to be subjected to an atmosphere other than Earth's since Viking's twenty-one years earlier, held. The parachute deployed and didn't tear. The braking rockets, a simple, solid-fueled design not unlike skyrockets, slowed the lander at the last minute as intended. And then came the moment of truth—at the proper moment another small rocket engine fired, inflating the seventeen-foot-high pyramid of Vectran airbags ten seconds before Pathfinder reached the surface. . . .

And the assembly bounced, and bounced. Then it bounced some more. After hitting the ground at 45 mph, faster than any spacecraft had ever encountered a planetary surface and survived, it bounced at least fifteen times in all. Then it rolled to a stop. It was 3:00 a.m. local Martian time.

After time for checkouts, preparations, and a bit of celebration in mission control, the triangular pedals lowered from the base of the landing stage. As they did so, they righted the machine to the proper orientation. There it sat for the remainder of its first cold night on Mars, marking the Fourth of July holiday and the twenty-first anniversary of the landing of Viking 1. America's first roving machine was on Mars.

Manning recalls landing day: "I was the flight director during the landing, which means that I knew when everything was supposed to happen. We had a transmission delay of about eleven minutes [due to the distance between Earth and Mars], so by the time we heard anything, it had either worked or it had not. Of course, we had practiced this extensively with our simulation setup [next to the sandbox]. We put a simulated eleven-minute delay into the signals between what was going on in the test area and the control room. So even though we could talk from one room to another in seconds, it was like adding 125 million miles of distance. We had practiced so much that when we actually landed, it felt fake!"

The next day, the little rover detached itself from its restraints and, slowly and daintily, crawled down the ramp. It managed to avoid snagging on the deflated airbags, which was another area of critical concern, and rolled the first two of six wheels onto Mars.

Pathfinder had landed in an area called Ares Vallis, a region chosen using twenty-year-old Viking orbital imagery. MER's and Curiosity's landing sites would be picked with far more precision due to the ultra-high-resolution orbital imaging that came later. Luckily, Ares Vallis turned out to be a reasonably good combination of a nice, flat, and safe surface that was still interesting enough to titillate the geologists. The region had been the recipient of vast water flows in the distant past, promising a fine variety of rocks and soils to examine.

The Sojourner rover spent almost three months driving around its small backyard. It never ranged much beyond thirty feet from the lander, as its radio transmitter had a limited range. In addition, they wanted to keep an eye on the little rover from the lander's cameras. Its first target, a rock affectionately dubbed "Barnacle Bill," was a scant fifteen inches from where the rover first drove onto the planet's surface. Many more specimens would follow, with such names as Yogi and Wedge. There was even a nice little collection of targets they called Rock Garden.

While this was occurring, the lander took its own photos, the first color panoramics of Mars since 1976. The images were stunning.

Everything seemed to be going swimmingly for the first six weeks, and then something happened that would become a regular occurrence as modern computers (albeit creaky ones by today's standards) met the red planet. The lander decided to attempt a reboot. It didn't ask, nor did it warn anyone. It just decided to. It took a day to fix the problem. This is one downside of modern, reprogrammable flight computers—they have occasional hiccups. Subsequent rovers have suffered similar challenges.

Upon regaining communications, the controllers noticed that the rover was not moving. It had stopped, jammed up against the aforementioned rock named Wedge. The tilt sensor had told it to stop before it hurt itself, and there it sat, awaiting instructions from home. After an intense conference, they radioed instructions Marsward and the machine freed itself, continuing its slow trek. Progress was measured in inches or a few feet per day.

Sojourner investigated more rocks with its instruments, basic versions of the much more sophisticated units that flew on MER and Curiosity. Its APXS—for Alpha Proton X-ray Spectrometer—visited many rock targets, sniffing out their composition with its limited sensing power. It could take days to evaluate one rock, but they were learning all the while. The rover's three little cameras, two on the front and one on the back, guided it and provided ground-truth images that were relayed to the lander and sent home along with the lander's own images, tracking the rover's progress. They tried out various experiments with the drive system—locking one side and turning with the other, or locking five wheels and allowing the sixth the dig a trench to observe the soil underneath. The lessons learned would guide rover design for a decade.

Pathfinder also reported Martian weather, the first meteorological reports from the planet in twenty years. The average day temperatures were below twenty degrees Fahrenheit, and the nights plunged to −104.

The rover had been designed to last one week and the lander a month. But in a pattern that would become synonymous with JPL's missions, both were still going strong over two months later. By early October, however, the computer-reset and communications problems were becoming more frequent, and it was clear that the duo's days (or, to use Martian terminology, sols) were numbered. By September 27, communications were failing, and the lander breathed its last on October 7. The surface exploration had lasted almost three months.

Fig. 6.2. HELLO, YOGI: In 1997, the Mars Pathfinder rover Sojourner spent almost three months exploring a small area near the lander. As seen here, it is investigating a rock they named Yogi (yes, after the cartoon bear) with the APXS instrument. *Image from NASA/JPL-Caltech.*

In its brief life, Pathfinder and Sojourner examined sixteen rocks and returned seventeen thousand images. The first on-ground evidence of water in Mars's distant past was found, though the area seemed to have been dry for at least two billion years. The rocks were volcanic in origin, demonstrating a geologically busy past for the planet—long suspected but a welcome confirmation. Lots of other scientific data were returned. But the most important thing was what they learned about going to Mars and how to land, drive, and survive in the harsh Martian environment.

It would all pay off in just a few years when the Mars Exploration Rovers headed off to the red planet.

CHAPTER 7

CURIOSITY'S COUSINS: SPIRIT AND OPPORTUNITY

Pathfinder had outstripped even optimistic projections. The concept of a roving vehicle, delivered directly to Mars without settling into a parking orbit first, worked brilliantly, and on a budget.

What was called for now was an ambitious follow-up: larger, more capable rovers to traverse far-greater distances and explore more targets in a much more detailed manner. But the year was 2000, and it was not yet time for Curiosity. An intermediary design took the form of the Mars Exploration Rovers. The MER rovers were extremely successful and have gone far past anyone's wildest dreams. The accomplishments of these machines, especially the still-operating Opportunity, nearly defy belief.

But as these roving machines were being built, JPL got a black eye. First one, then a second mission failed in 1999. The Mars Climate Orbiter (MCO) had sped off to Mars in December 1998, arriving there in September 1999. Intended to aerobrake into a Martian orbit—that is, enter the atmosphere just enough to slow it but not enough to make it crash—the unfortunate spacecraft plunged into the Martian air at the wrong altitude and augered in. It the end, the embarrassing revelation was that somewhere along the development path of the software, there had been a failure to convert from English units to metric ones. The altitude settings were way too low. Oops. In a business where a few milliseconds or a fraction of a degree of trajectory error can cause calamity, this was huge—and damn embarrassing.

On top of this, a second JPL mission, the Mars Polar Lander (MPL), failed in December of the same year. When it arrived at the red planet, the spacecraft retrobraked into its descent, ejected its parachute, fired its braking rockets—and went promptly and permanently off the air. It just vanished. No indication of landing, no

63

final call for help, just silence. What actually killed the mission is not entirely clear, but it is suspected that one of the violent events that occur during a landing—firing of braking rockets, deployment of the landing gear, *something*—had caused the software to think it had reached the surface safe and sound. The computer shut off the rocket engines and the machine fell from perhaps 120 feet or so—too far to survive the impact. It was like a skydiver cutting loose his parachute at a couple hundred feet, and the results were similar. It died.

JPL was zero for two with Mars in 1999. A lot of hope was being placed on MER's shoulders, and if for some reason this mission was unsuccessful, there would likely be no MSL. Those involved knew the stakes and rose to the occasion. They would absolutely not fail.

The new millennium brought renewed hope even as it fostered seemingly endless and painful reviews. JPL's managers wanted to know what had gone wrong in the months previous, and atop them sat the NASA juggernaut, also waiting to know what had happened to its hundreds of millions of dollars. And atop *them* . . . some members of Congress, looking for someone to blame, or for excuses to further trim the space budget.

Meanwhile, planetary science does not stand still . . . missions ready to go were allowed to continue. The Mars Odyssey orbiter reached the red planet in October 2001 and began returning spectacular results. It was fortunate for all concerned that it did.

Throughout these trials and (more recent) successes, the Mars Exploration Rovers were being readied. The MER rovers, Spirit and Opportunity, were scaled up by an order of magnitude in every way over Pathfinder. Weighing well over ten times as much as the Sojourner rover at 410 pounds, with much more robust instrument packages, they would assault Mars as twins. Not since Viking had a mission to Mars had the luxury of a backup in case of single failure, nor are we likely to see it again. It has simply become too expensive.

There was a new and unique twist to this mission besides the size and complexity. The twin rovers would be able to regularly use NASA's Mars orbiters, Mars Odyssey and the older Mars Global Surveyor, to relay messages back to Earth. This gave them a much wider time slot in which to talk to JPL and far more bandwidth than surface antennae such as that on Pathfinder. This blending of assets served MER perfectly and was a good training ground for the future MSL, then being planned. MSL would have much more demanding data-relay needs, so success in this with MER was crucial.

As before, the machines would be powered by solar panels, except that rather than sitting as a small rectangle atop the microwave-sized Sojourner, these folded out like beetle wings, giving the rovers a decidedly insectoid look. Aboard were larger and far more sophisticated instrument packages, but as with the rest of this program, they were an evolution of what had gone before. It was a good case of trying, learning, and improving with each mission.

Steven Squyres, a professor at Cornell University, was the principal investigator, in effect the science boss, of the Mars Exploration Rover mission. Tall and lanky, he brims with the confidence that only a lifetime spent as an astronomer and a decade of Mars-rover operations can inspire. He speaks plainly and easily, has a breezy sense of humor, and is the kind of guy you'd want as a geology (or calculus or chemistry) professor. MER, he said, had a far more ambitious plan than did Pathfinder, now that they had the basics licked. "The primary goal of MER was to go to two places on the Martian surface and try to learn what conditions were like in the past, and then discern if they might have been habitable." He paused. "You know, Mars is a cold, dry, and desolate place, but in the past conditions were probably different. So we tried to choose two places that looked, from orbit, not just to be good places to land on, but that appeared to have traces of water in the past. We hoped to really read that story in the rocks and see how habitable it might have been."

His distinctive leadership of the mission can be sensed when he discussed the instrumentation of the rovers: "The way I chose the payload was by picking a set of tools that were as capable as possible. It was sort of like trying to design a Swiss army knife, finding the most capable tools you can and putting them on the vehicle. There were also some very serious practical considerations regarding availability of technology, its maturity, and the risks of the technology you would buy. Everything has got to work. As things have turned out, I'm quite happy with the payload that I chose."

While the MER rovers' level of sophistication pales when compared to MSL, the machines were still stuffed with everything the planners could afford to fly. The most obvious was the camera mounted atop a tall, folding mast. Called Pancam, it offered a splendid, wide field view of the terrain ahead of the rover. Navigational cameras offered an even wider view for the rover drivers. A small camera mounted on the end of the robotic arm, the Microscopic Imager, allowed for close-up views of rocks and soil. Four more small cameras completed the collection, these also for driving and hazard avoidance. All these imagers would be refined and augmented for MSL.

A Thermal Emission Spectrometer was mounted in such a way to see the rocks and soil ahead and help to identify promising areas for investigation. By observing the rocks in infrared, the device would be able to effectively see through the dust that clung to everything on Mars and also sense how quickly heat escaped over time. This would aid in determining the possible composition of a rock or outcrop.

Out on the rover's arm, along with the microscopic camera, were four more tools. There was another spectrometer, this one designed to be held up close to interesting rocks. The Alpha Particle X-ray Spectrometer, or APXS instrument, would bombard targets with energy and observe the resulting reaction. This was a more-sophisticated version of what had already flown on Pathfinder and would in turn be improved for Curiosity. There were magnets to gather ferrous soil, and finally the Rock Abrasion Tool or RAT, a wire brush for cleaning off dusty rocks prior to examination.

Deep in the chassis was the onboard computer similar to the one that had given Pathfinder's Sojourner rover such fits. This little CPU, an IBM RAD6000 in the jargon, was way out of date when compared to those then being used in consumer computers on Earth, but it was the most affordable unit that had been "hardened" against radiation and was of military quality. Similar units had powered Macintosh computers back in the early 1990s; now it would go to Mars. It ran at a now unimaginably slow 20 megahertz and was supported by 128 megabytes of RAM. Today it would cower in a corner if it met a modern smartphone. But the thing was designed to withstand a nuclear attack, which should make Mars child's play by comparison.

Another evolutionary element of the MER rovers was mission management, which was in many ways a very scaled-up version Pathfinder's. While MER was an entirely different beast than Pathfinder, it had actually grown out of simpler origins. In a set of circumstances that will surely never be repeated at JPL, the entire mission—both rovers and all the science conducted by them for over a decade—fell under one man. Steve Squyres had been tapped early on for the mission. His pedigree was a profound one: he had worked on Voyager (as a very young man), the

Magellan Venus mission, the NEAR–Shoemaker (Near Earth Asteroid Rendezvous–Shoemaker) mission, the Cassini Saturn probe, and a European Mars orbiter called Mars Express.

When Squyres was assigned to MER, however, the mission had been planned as a single lander. Then it morphed into a rover. . . . Then two rovers. It just grew and grew, and soon it had 170 scientists working on it, all for one principal investigator—Squyres. It was a busy time that included his scientific work on the aforementioned Mars Express as well. As Grotzinger, who also worked on the MER mission, once commented, Squyres was often like "a flea on a hot griddle." He seemed to be everywhere and involved in everything. Of course, Grotzinger would soon find himself in similar circumstances.

During mission planning it had all seemed doable. After all, each MER rover was supposed to have a ninety-day primary mission, and, if they were lucky, the machines would survive a Martian winter and soldier on for a year. If they were lucky. Nobody at the time dared suggest that one of the rovers would still be busily driving along the Martian surface a decade later.

But first they had to get the rovers to Mars.

After a multimonth trek, and in a repeat of the techniques pioneered by Pathfinder, these two comparative behemoths would hurtle directly into the Martian atmosphere upon arrival. Within a few seconds, the heat shields would go from space-cold to incandescence as they slammed into the thin air high above the planet. Parachutes would plume open, braking rockets would fire, and in the bizarre ballet pioneered by Pathfinder, the tumorous beach-ball-like airbag cluster would inflate instantaneously when the rovers neared the ground.

Upon impact, the spacecraft would bounce, nearly one hundred feet high and many times, rattling and shaking the rovers ensconced within. Yet within minutes they would both be stationary, with the airbags deflated and lying across them like dead skin.

In order to use the airbag system, they had to first slow the spacecraft from its interplanetary cruise from Earth. This would require a parachute. But Pathfinder had weighed a lot less than MER, and the Vikings had orbited Mars before landing (which slowed them down). Like Pathfinder before it, the MER rovers would be coming straight from Earth, shot like artillery shells leaving Florida and aimed at a spot over a hundred million miles distant. The speed at which they would arrive at

Mars was very high and would require a new parachute-system design . . . and we already know how difficult developing parachutes for Mars can be.

Fig. 7.1. BOUNCING TO MARS: Like Pathfinder before them, the twin Mars Exploration Rovers arrived encased in inflated cloth bags that looked, and acted, like giant beach balls. Each rover rebounded over a dozen times but eventually landed safely. This landing system was not strong or accurate enough for Curiosity. *Image from NASA/JPL-Caltech.*

Rob Manning had discussed the parachutes with me at some length, but I'll just repeat a bit of it here. He detailed the gruesome process of testing the parachutes for MER, as (in a repeat of Pathfinder's teething pains) they ripped, snarled their lines, and ripped again. "We were still testing the parachute in the wind tunnel with less than a year to launch, and still having failures. We attached big weights to the parachute and threw it out of a helicopter. Sometimes the parachute opened late, and at that point it was going so fast that it just ripped open. So we would retime the opening and it would still rip. The whole setup was just really hard to get right." They tested and tested, until they felt they had wrestled the problems into compliance. But right up until landing, the parachutes were still a concern.

These and a thousand other challenges were ground down to size and beaten into submission, and the twin rovers launched in 2003, arriving at Mars in 2004.

They entered the Martian atmosphere at high speed as anticipated, the parachutes deployed and inflated as planned, and the airbags inflated on cue and cushioned the rovers as they bounced to a stop. Of course, all this happened a hundred million miles away on Mars, and the engineers had to wait over fifteen minutes for the confirmation signals to reach Earth. But when they came in, everything had worked damn near perfectly and each rover landed in the region at which it had been aimed. The landings were engineering tours de force.

As it turned out, the area in which Spirit descended, a crater named Gusev, was a flat, billion-year-old basaltic plain. This was interesting in its own right but not what the scientists had hoped for—the basalt resulted from lava flows that obscured the older sedimentary material beneath. The rover would drive and explore, but its primary objective—finding evidence of a watery past and a possibly habitable environment—would be a challenge. But then Opportunity landed.

This second rover came to rest in a region called Meridiani, which was much more geologically varied than Spirit's. And better still, Opportunity rolled to a stop inside a crater. Named Eagle Crater, it was in effect like a huge drill hole deep into the Martian surface, created by nature. Opportunity was the benefactor. "We lucked out!" Squyres enthused, "We discovered that we were in a giant impact crater that had all the things that we could have wanted exposed there in the wall of the crater. In two months, much of the important science was revealed to us."

The discoveries began to pour in: they had direct and verifiable evidence of copious amounts of water on Mars in the past. This was represented in the sedimentary layers found throughout Meridiani and was also evidenced by the small hematite spherules, dubbed "blueberries," found everywhere the rover went. Grotzinger was a prominent member of the MER team, and this discovery just whetted his appetite for more.

"I was the only sedimentologist on that mission," he recalls, "but there were other geologists who were familiar with sedimentary rocks. When we rolled off the lander and went to the first outcrop and looked at it, we said 'Yeah, these are sedimentary rocks.' Then the question was, do they have something to do with water? Soon we were seeing the 'blueberries,' the spherules, lying all over the place, and we saw that they were coming out of the outcrop. At that point, we labored night and day and debated what the various origins might be. The berries were a big part of it. One option could be that they were volcanic; the other option could be that they were produced by a big impact. The third option, which was considered to be the most exotic, was that they were sedimentary. We worked through this, and it was a lot of effort," he remembered with a smile. "It was about three weeks. Seventy-five percent of the team was pretty well convinced that they were sedimentary in origin."

Fig. 7.2. ROVING ENDURANCE: This composite portrait of Opportunity driving at Endurance Crater shows the ubiquitous "blueberries" on the ground. The MER rovers pioneered much of the technology, especially the autonomous driving techniques, used for Curiosity. Opportunity's mission continues today. *Image from NASA/JPL-Caltech.*

For Grotzinger it was a turning point. "For the next seven years [and] when we got to Endeavor Crater that's all we saw." But one can take nothing for granted on Mars. "We decided to repeat the experiment. Every time we went to a new outcrop, we thought, 'This is Mars, everything can totally change!' But then we get to another outcrop and we say 'Hmmm . . . here is the same thing.' But it wasn't *exactly* the same thing. Sometimes the environment that it indicated was wetter, and at times it was dryer; in some cases, we had fields of ancient sand dunes that were frozen in time." These are subtle differences, but they make a lot of difference in geological terms, especially on a currently arid world like Mars.

The evidence of sedimentation was clear. They had seen cross-lamination, a sure indicator of water-deposited sediments. And these layers were deposited not by a lazy pool of standing water but by a flowing stream or river. Mars had not just been damp but had at one point been drenched in water.

More blueberries were found—they turned out to be quite common in Meridiani. One area was so densely adorned with the blueberries that they dubbed it "Berry Bowl." Here at last were enough of them clustered in one spot for Opportunity's spectrometer to finally get a solid reading (individual berries are very small). They checked out as hematite—a water-formed mineral.

The presence of water was becoming almost boring . . . to the public. But not to the scientists working on MER.

"The whole theme of this Mars exploration program has been to follow the water, to understand the possibility of life on Mars. Clearly all life that we know of needs water in its cells and in its environment to survive, so it's always been the major goal of these Mars missions to understand the role of water in both the history of Mars and also the evolution of Mars," said Squyres.

"The water was primarily beneath the ground," he mentioned about the period when Opportunity was exploring a crater called Endeavor. "It was not nice, pure water, it was salty stuff, and it was probably very acidic." That was not good for living things. But this was merely sauce for the goose to the planetary geologists. It would be one more thing to add to Curiosity's wish list—evidence of benign water in epochs past.

Across Mars, Spirit would not have it so easy. When the rover began sniffing the rocks in the Gusev area, one of the early discoveries was a mineral called olivine. It was not something they wanted to find, as it indicated a long-dry environment.

"Olivine is a mineral that tends to be present in unaltered igneous rocks," said Squyres, "so finding it was a disappointment because that was one of the first things that made us realize that instead of landing on sedimentary rocks, we landed on a lava flow, one at least a billion years old. It took a while to sink in what we were dealing with, that the sediments we were looking for were completely buried in the lava." He sighs at the memory. "Once we finally realized that, we had to move somewhere else." That's when Spirit drove on to explore the Columbia Hills.

Once there, they had more luck. These rocks had been altered by water and were not just an endless expanse of flat basalt. As the amount of olivine declined, they found more indicators of a wet past, especially sulfates, which are indicative of clays and therefore water. Olivine decomposes in moist environments, so they were headed in the right direction.

The list of discoveries by the MER rovers can fill its own book; suffice it to say here that the list was long and at nearly every turn it led to the conclusion that Mars had once been very wet. This in turn dictated an even tighter focus for the mission of Curiosity: further refine the likelihood of an ancient habitable environment.

In July 2007, vast dust storms began swirling around the planet. With no oceans to intercept them, Martian dust storms become global events. The daytime illumination got dimmer and dimmer, and the rovers, which were dependent on solar power to provide electricity, were running down. The engineers began to worry; at one point, the sunlight was reduced by 99 percent. On July 20, 2007, NASA released a terse statement: "We're rooting for our rovers to survive these storms, but they were never designed for conditions this intense." The rovers were parked to wait out the storm, but if the power fell too low for too long, it would be fatal. They could switch to low-power fault mode but might not have enough energy to come back when the skies cleared. But clear they did, and by August 21 it was sunny enough to recharge the drained batteries and for driving to begin anew.

The use of solar panels for power had just demonstrated one of its disadvantages. Between the regular coatings of dust and the fact that the sun is fairly weak when you get as far from it as Mars is, a different source would be needed for Curiosity. The complexity of the new rover would demand one more reliable and consistent, that could not just power the rover through its duties but also help to supply current and heat through the long, cold nights.

Note #1 to metaphorical self: *Find better, continuous source of power.*

By early 2009, echoes of the Pathfinder computer glitches were reverberating through Spirit's brain. The onboard flash memory, not dissimilar from the thumb drive you might use to back up photos or transfer files to your own computer, began to behave erratically. It would download instructions and then fail to execute them or neglect to upload all the data gathered on a given day. It seemed to have developed a mind of its own. Then, like a tired dog curling up for a nap, it began to reboot spontaneously. These issues would crop up for the remainder of the rover's life.

Note #2 to metaphorical self: *Supply a backup computer system.*

Spirit managed to drive almost five miles across Mars in its six years of operation. Then, just past the five-year mark, it became bogged down in sand. And there Spirit sat, effectively functioning like a stationary lander, until the radio failed in March 2010. JPL worked on the problem until May 2011 and then, after a herculean effort, called it quits. They held a formal farewell ceremony for the rover on Memorial Day.

But very little is wasted on Mars. The long hours put into the attempted recovery of mobility—planning, software simulations, then mechanical simulations with the rover's earthbound twin in sand both at JPL and in the California desert—would be very educational. They would use this education for the continuing mission of Opportunity and in the ongoing preparations for the Mars Science Laboratory.

Note #3 to metaphorical self: *Use larger wheels that spread the weight of the vehicle across a wider surface and have lots of traction.*

Opportunity has had a better go of it. Now nearing its tenth anniversary on Mars, it has outlasted its three-month mission exponentially. In its long life, it discovered the blueberries scattered across Meridiani, located the first meteorite ever found on another planet, drove deep into vast craters, and studied the strata found there, which was a geological gift from the planet-drilling impact eons before.

Opportunity's long life and extended driving has allowed engineers to refine a technique they had only experimented with on the Pathfinder mission: autonomous driving. For much of the twenty-four-hour, forty-minute Martian day, the rovers were out of contact with Earth. They could either sit and wait until they could get live (albeit delayed) driving instructions, or they could learn to navigate on their own. This feature had been built in from the beginning, and the decade of operations gave the rover drivers precious experience in autonomous hazard avoidance and advance planning. Yesterday's images would be used to plan tomorrow's drive, or the next day's. A wonderful software program called RSVP—short for Rover Sequencing and Visualization Program—was developed. This tool allowed mission planners to download images from the rover, map them into a 3D model of the surface, then select routes to drive to specific targets. But that was not the amazing part. Similar software, living in the rover's limited onboard computer, allowed it to make its own surface maps. With some advance direction from the ground, the machine could then pick its way through the landscape, slowly at first, then with increasing confidence. Of course, it worked better in flat, open plains than it did in challenging, rocky, or cratered environments. But both MER rovers had plenty of flat terrain to cover, and the experience taught the JPL engineers a lot about autonomous driving. Pathfinder pioneered it, but MER matured it. Curiosity would perfect it.

Note #4 to metaphorical self: *Apply more autonomy, both in driving and in surface operations.*

In a related evolution, Opportunity's long and storied exploration of Mars has paid off in another, less direct way. With the fleet of orbiters that continue to circle Mars, providing ongoing maps of the surface far below, experience with the MER rovers has given MSL mission planners a far better understanding of how to correlate orbital imagery with what is actually on the ground. This would pay vast dividends when selecting a landing site for Curiosity . . . though it did little to dampen the passions, or temper the debate, about exactly where to set down.

Note #5 to metaphorical self: *Having a pair of orbiters to relay information and provide updated surface maps is an incredible asset.*

One final note: on May 16, 2013, Opportunity, having then driven 22.2 miles, officially broke NASA's off-world driving record set by the Apollo 17 lunar rover. That machine was piloted across the moon's surface by astronauts Gene Cernan and Harrison Schmitt, and covered twenty-two miles of hard, undulating lunar terrain. It did this during a three-day lunar stay, however, as compared to the Mars rover's decade. So, while Opportunity's record stands, there is still something to be said for human exploration. Once people are there, things will move a *lot* faster. (Yes, that is my hand you see raised in the corner, I am volunteering . . .)

Opportunity continues to rove the regions around Endeavor Crater. There have been ongoing problems with its aging robotic arm, and the motor driving its front right wheel has been using too much electricity for years. In fact, they have had to drive backward to compensate, and have done so for longer than they drove it as designed, arm forward. But the golf-cart-sized rover shows no signs of succumbing to the harsh Martian environment anytime soon and continues its slow but steady explorations to this day.

Note #6 to metaphorical self: *How cool will it be to have two rovers working on Mars again once Curiosity arrives?*

Lessons learned, time to step it up—a lot.

CHAPTER 8

A WHOLE NEW BALL GAME

E ven before the MER rovers landed, in fact, while they were still being built, JPL was tasked to come up with the next great thing. Based on what they hoped MER would achieve, they would spec out and design a new mission architecture that would go far, far beyond what even MER could do.

It was expected that much of what was learned with Pathfinder, and later with MER, would be applied to this new mission. And indeed, much was. But this new machine was so grand in its scope, so massive in its design, that much of what had gone before was not terribly informative. JPL'ers didn't just think outside the box, they threw it away.

In the late 1990s, a team of engineers, scientists, and managers were assembled to begin the process of identifying the shape and goals of such a new program. They were given a blue-sky mandate: think big and wow us. The result became known as the Mars Smart Lander, which would eventually morph into Mars Science Laboratory. Of course, blue-sky is fine until the price tag comes in, then budgets tend to reign in ambition.

What was known up front was that there was a keen desire to put ever-increasing levels of capability into the science packages on the rovers. The MER machines, Spirit and Opportunity, were already a quantum leap beyond Pathfinder's Sojourner. The MSL would be another huge step, but in exactly what direction and to what ends needed to be defined.

Overall, however, with over twenty years of technological advancement since the time of Viking, Curiosity would make the 1976 mission look like it had the capabilities of a sewing machine. The new rover would be more tightly focused and supplied with vastly better instrumentation than anything that had gone before, while building on past accomplishments.

Pathfinder had proved the concept of a roving vehicle that was delivered directly to Mars without settling into a parking orbit first. It also pioneered the a robotic arm and associated instrumentation, along with its innovative suspension system.

The Mars Exploration Rovers were in turn wildly successful and have gone far past NASA's wildest dreams in terms of accomplishment and longevity. The utility of improved instrumentation and the usefulness of orbiters for data relay were validated. Increased autonomy for future rovers was demonstrated.

Planetary-exploration programs build on experience and knowledge, and NASA is very good at incorporating lessons learned. JPL has made it into a science. While MSL had a lineage going back to Viking, it would not be tasked to look for microbial life as Viking had—that had turned out to be a far more complex undertaking than was understood back in the 1960s. In the time since Viking's mission goals were defined, truly weird life-forms have been discovered on Earth, such as the extremophiles, exotic critters that exist in hot, lightless environments like the "black smoker" geothermal vents on the deepest ocean floors. If things this odd, unexpected, and unusual could live on Earth, there was no telling what forms they might take on Mars. Other examples are living organisms found in places such as the Antarctic, that live *inside* rocks (near their surface), protected from harsh conditions but still able to draw sustenance from the environment. It appears that life is far more tenacious and clever than we had thought in the 1950s and 1960s, so another life-science mission to Mars would have to wait until habitable environments had been identified there. We would need good evidence to believe that life might currently survive on Mars to embark on such a costly and ambitious mission.

Once MSL's mission goals were defined (and constrained), this helped to inform choices for instrumentation to put on board (though there would still be plenty of debate over the final selections). The tools on Pathfinder had been minimal, and Spirit and Opportunity had expand on these greatly. However, MSL specifications were being planned well before the MER rovers landed and began operations, so decisions would have to be made largely without MER data and then would have to be refined later.

If you are beginning to sense nonideal circumstances here, I'll know you have been paying attention. Due to even this moderate pace of the Mars exploration program, two-year departure windows and the vagaries of congressional funding, planning for a current mission was usually based on data from missions two generations previous. So decisions about how and where to land Curiosity, and what to put on board, were hinging on what had been learned up to and including Pathfinder. Then, as data streamed in from the current mission, MER, they would be incorporated into MSL's plans. It's a system that has evolved to be tolerant of flux but can occasionally hold some surprises and make for some difficult choices. Fortunately, the 1990s and early 2000s were such a rich period for Mars orbital missions that the landscapes have been mapped at very high resolutions, with even small features visible, so mission planning for the future has become much more of an exact science.

Back to instrumentation. Pathfinder's lander had carried cameras with a spectroscopic ability. The Sojourner rover had cameras on board as well as the Alpha Proton X-ray Spectrometer (APXS) on the end of its arm, which could bombard target rocks with high-energy particles and read the resulting spectra.

The MER rovers' onboard cameras were much higher resolution and far more capable. A variety of spectrometers, both active (like Sojourner's APXS) and passive, provided a huge improvement in the ability to understand the composition and nature of rocks and soil. The RAT (Rock Abrasion Tool), a rotary wire brush, allowed for the cleaning of rocks before they were investigated. A microscopic camera on the arm gave highly detailed close-ups of chosen targets. But as impressive as these improvements were, they paled next to what was planned for Curiosity. If one looks at the lineage from Sojourner through MER, it is a bit like comparing a little red wagon with a camera to a Jeep with a few scientific tools mounted to the front grille. Moving on to Curiosity equated to a school bus crammed with a full-on science lab. This was the challenge the JPL engineers, as driven by the scientists, placed before themselves. We will look at the onboard instrumentation for Curiosity soon. But first: where to land the massive rover?

The designers would build on data from previous rovers as well as the vast trove of new images and data coming in from the orbiters. By following the trail of ancient water features and geological evidence of the same, Spirit and Opportunity would provide data to better interpret the high-resolution orbital images and, by extension, help to determine MSL's landing zone.

Each Mars landing has been more precise than its predecessor. In the time of Viking, the safety of the lander had been of paramount concern, with interesting surface geology a distant second. When the landing area for that mission had been planned in the 1960s and early 1970s, the images available to do that had all come from Mariner and Earth-based telescopes. As we know, the detail from telescopic operation was pitifully low—even continent-sized areas were fuzzy. When the Viking planners were considering landing sites, much of their available knowledge came from Mariner 4, which had mapped only *1 percent* of the Martian surface— hardly enough to base a decision on. Mariners 6 and 7 had improved this, but they were still flyby missions that imaged only a swath of the planet as they dashed past. You could have flown over a Martian city and not known what it was.

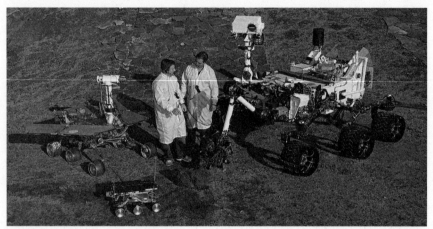

Fig. 8.1. MEET THE FAMILY: For a sense of scale, JPL posed the three generations of Mars rovers together (along with a couple of willing volunteers in the ubiquitous lab coats). Pathfinder's Sojourner (1997) is to the lower left, a Mars Exploration Rover (2004) is to the upper left, and, dwarfing everything, Curiosity is to the right. The radioactive power source has not been attached yet. *Image from NASA/JPL-Caltech.*

By the time of Mariner 9 in 1971, image resolution was about 1,100 yards— anything smaller than that would be invisible. And yet it would take just one crater rim or one rock more than about eighteen inches high to destroy the lander. It seemed like a hopeless task, but with the momentum of the space race pushing them, they proceeded.

Of course, there was more to it than just the pictures. By observing areas with spectroscopes and looking for moisture and mapping surface temperatures, much could be inferred about surface characteristics that could not actually been seen with the eye.

The Viking orbiters improved image resolution, but we are still talking about twenty-five feet for the smallest surface features.

Flash-forward to the 1990s. The Mars Global Surveyor, a replacement for the failed Mars Observer, arrived in 1997 and began mapping the surface for the first time since Viking. A lot of surprises were in store. Objects as small as about five feet were now visible.

Additionally, scientists at JPL would compare these pictures to what were called "context images," which were a wider view and showed the type of landscape the close-ups were located in. Between these two, researchers could infer a lot about the nature of the terrain. Of course, this was no help for the Pathfinder lander, which was arriving at Mars in the same year, but it was a tremendous help in planning for the MER rovers.

A reminder about Spirit and Opportunity: these rovers were expected to roll to a landing in areas covering seven by fifty-four miles. So even if you knew of specific objects within that landing ellipse that were potential rover killers, it would be a matter of probability whether or not those objects would interfere with the landing. The good news was that by using the airbag landing technique proved on Pathfinder, you could bounce and roll across the worst of them, and odds were that the machine would end up somewhere that was adequate to allow the rover to roll off the landing stage and proceed with the mission.

Curiosity would approach the forbidding surface in an entirely different manner, and one that entailed its own gruesome challenges.

One more major improvement in MSL's design was the way in which the rover would be powered. As I mentioned, Sojourner and both MER rovers used solar panels on their backs to gather the weak sunshine on Mars and convert it to electricity. This was routed to batteries and stored for later use. It's a great system in theory and can operate at least a decade as Opportunity has shown.

But solar panels have their drawbacks. One major limitation is that that they work only in direct sunshine. It's hard enough to get a lot of power out of these things on Earth, which receives four times as much sunlight as does the surface of the red planet. Solar panels on Mars have to work even harder than earthbound panels to generate much power.

Then there is the dust that is ubiquitous on Mars. Due to the geology and weathering processes of the planet, the place is lousy with talc-fine dust that the winds can accelerate to very high speeds and cause planet-girdling storms. Even lesser weather patterns carry thousands of tons of dust and dirt aloft in the wind.

And as with the 2007 dust storm that threatened the MER rovers, these events can be rover killers. The dust storms darken the skies and can end up coating the solar panels with a layer of grime that reduces their ability to receive light.

To overcome these limitations for Curiosity, the designers turned to a technology that, in spaceflight terms, dated back to the 1960s. In more brutal terms, it dated back fifteen years earlier. They would use nuclear fuel—plutonium—the very same stuff that incinerated Nagasaki in 1945 (the Hiroshima bomb carried uranium, a less-powerful nuclear source but still very destructive when used in a bomb).

Plutonium fuel had been used on the Voyager spacecraft that traveled out to the distant planets of the solar system—Jupiter, Saturn, Uranus, and Neptune—where there is almost no sunlight. Plutonium was also used to fuel the lunar-surface science packages left behind by Apollo 12 through Apollo 17, as well as the Viking landers. Space budgets of the time were vast, and there was a lot of plutonium around from the 1950s and 1960s (guess why?).

The device that uses this fuel for spacecraft power generation is called an RTG or Radioisotopic Thermal Generator (there are other acronyms used, but this one is the most descriptive). The "hot" nuclear fuel is placed inside a metal cask, and its slow radioactive decay causes heat. Surrounding this are thermocouples, metal strips that convert heat into electricity. So, in this use, the plutonium always remains subcritical, that is, it is not used for fission (atom-bomb) or fusion (hydrogen-bomb) reactions. It just sits and quietly gives off heat.

The upside is a nearly permanent source of power that is immune to dust storms, nightfall, or winter darkness. The half-life of plutonium is fourteen years, and it can provide power far longer than that. The Voyagers have been traveling the solar system for over thirty-five years and, while their power supplies are declining, they are still operating, albeit in a reduced capacity.

If there is a downside, it's primarily theoretical. Some people get upset when NASA launches nuclear generators into space—what if the rocket blows up? Won't it poison the atmosphere? Well, NASA's civil servants don't want to die a lingering, horrible death of radiation poisoning any more than you do. The plutonium is encased inside a containment vessel that is tested to be much stronger than is required for it to withstand an explosion and atmospheric reentry. If that event did occur, the cask-encased nuclear fuel would fall into the ocean, sink to the seafloor, and slowly radiate heat for about forty years. Of course, it's toxic much longer than that, but nuclear plants on the surface of our planet are far more dangerous. It's all a matter of containment.

There is another problem however: the United States used up its available supplies of plutonium long ago, and making more is not a trivial thing. So rather than reopen the old bomb factories, we have been buying it from our former archenemies, the Russians. Irony upon irony—a substance created and first used in the rage of war to destroy entire cities is now purchased by Americans from an old enemy that made that same substance to turn the United States into a pool of molten goo. It's a far better use.

Truly a whole new ball game.

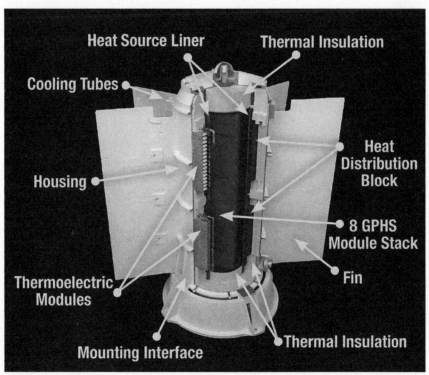

8.2. GOING NUCLEAR: A schematic of a Radioisotopic Thermal Generator, or RTG, similar to the one powering Curiosity. The gray rod at the center is the plutonium fuel, which on Curiosity weighs eleven pounds. The RTG is mounted upward at an angle on the back end of the rover. *Image from NASA/JPL-Caltech.*

CHAPTER 9

WHERE TO GO?

Remember the Mars exploration mantra: follow the water. The observations from Pathfinder regarding a once-wet Mars had been titillating, and those from the ongoing MER rovers were transformational. In the midst of this embarrassment of riches, the scientists had to decide where to send Curiosity. The process was far more public and open-ended than you might expect.

There is a tension between the mission scientists and the engineers—and it's all about the potential for wonderful discoveries versus arriving and operating with a margin of safety. In addition, on the science side, the potential for scientific return is paramount, and various locations on Mars were under discussion from that perspective. On the engineering side were people who loved a good challenge and would put the rover wherever they were instructed while warning about possible consequences. The scientists wanted an abundance of rock types, variations of surface terrain, new and ever-richer potential examples of water-altered rocks, and so forth. On top of both groups was NASA headquarters. They, of course, wanted great science, but please . . . *please* . . . don't lose the rover.

So, a bit of pressure here.

Grotzinger was at the helm by the time this discussion got serious in 2007. As with so much of the mission, the landing-site discussion involved many viewpoints that needed to be reconciled. A multitude of potential landing sites, all offering something wonderful, were being discussed. Each site had its promoters and adherents, and all views were held with great passion. Somehow, a final decision would have to be reached, and soon.

When interviewing Grotzinger about the process of narrowing down landing sites, I got a sense of great restraint at work in his retelling—the process must have been pretty tough. The scientific community was trying to decide where to drop their once-per-decade, $2.5 billion rover. It could not have been easy . . . or even always collegial. "There was a lot of discussion, some acrimonious. People had their views of what the planet should be. There was sort of a status-quo view that it was first and foremost a volcanic planet and anything that involves water is likely going to be a hydrothermal environment [i.e., one involving ancient hot-water processes]. That's a totally legitimate point of view. When you look at the geological context [and] some of the places that were advocated, it makes perfect sense. In other cases, like when Eberswalde Crater came along 2003, which was not accessible to MER in 2004 for the landing-site selection because the [landing] ellipse was too big, it came back for MSL." Eberswalde is a geologically interesting, partially buried impact crater. It would end up in second place as a preferred landing site for MSL.

"Holden Crater, right next to Eberswalde, also has a lot of evidence for what look like sedimentary deposits. Gale Crater had been considered in 2004, but nobody really knew what to make out of it. It's just kind of a weirdo. But more importantly, you couldn't fit a landing ellipse in there. Now everybody agrees that all these places have had water. The question became, which one of these would have been the most habitable environment? If microbes would have been there, people can imagine that there will be certain metabolic pathways where certain types of organisms could grow. Then there are other places where other types of microorganisms might have grown. What everybody was really trying to come up with was what they thought might be the best scenario. I realized that the trick for this mission was not whether life had evolved on Mars, it was that there could be places that could have been habitable environments. The trick was to find the evidence of it. The evidence isn't always preserved," Grotzinger concluded.

Now it got tricky. A decision would have to be made based on the likelihood of finding things once the rover was on the ground that could not readily be identified from orbit. It was the Viking problem all over again . . . one had to *infer*, using other observations and a sense of context. For, while the images from the Mars orbiters were better than ever, not everything can be seen from the orbital perch. It is tough, for example, to observe carbon deposits from orbit. "We can see carbonate minerals rarely, very rarely, in one or two places," Grotzinger added. But that in itself was not much to go on. "It was just a big uncertainty. So I basically said, 'Look, guys, we're just going to have to go there and hope that some of these things that we imagine are actually preserved.'"

There had already been two big meetings about this, called MSL landing workshops. Now, in 2007, they would have the third. "They had managed to pare down sixty targets to twenty. But it wasn't until the third workshop that people started to really work it because in a few years we're going to have to get down to one. It was a big challenge. The process is open, so anybody in the community can engage with it. So I arranged one part of the workshops for some people that had experience looking for habitable environments on the early Earth. We are looking for preservation of organic carbon."

A short list of candidates had been compiled and had to be evaluated on both scientific and technical merits. In the past, difficult terrain and the lower-resolution images from the orbiters (which forced more guesswork) would have combined with more inexact methods of landing the rovers (i.e., the airbag system) to eliminate some of the potential sites. But not this time.

Grotzinger continued: "I discussed it with the program scientist at NASA headquarters, and we realized that for the first time in the history of planetary sciences, the engineers weren't going to kill a single potential landing site." In a historical first, the folks handling the landing system and worrying about the mobility of Curiosity were not vetoing any of the top sites still standing. "All of our top four finalists were viable! Holden was viable, when it had never been viable before. Eberswalde was viable. Gale was viable, and this place called Mawrth [Vallis] was viable. *All* of them are viable. How the hell are we going to reduce it down to just one?" Holden was a vast crater that seemed to be an ancient lake, and Mawrth Vallis was an ancient outflow channel possibly littered with clay, also indicative of water and sedimentation.

What Grotzinger and his NASA colleagues decided was to present the rest of the scientists with a short list of choices. They were all reasonably safe and could be precisely targeted, given the accuracy of MSL's landing system.

NASA convened the working group again and tossed the options into the fray. "The community made a list of pros and cons and then people took a swipe at everybody else's chosen site," Grotzinger recalled. "Then we sat down and looked at our portfolio like an investor would, as a science team, which worked incredibly well." They pared the list down to four candidates, and took a vote. At last, they would have a winner.

Or perhaps not.

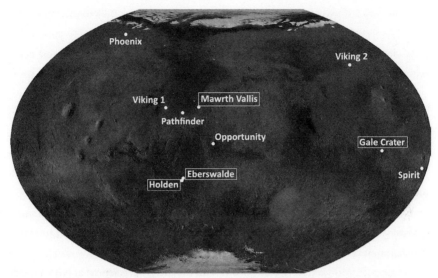

Fig. 9.1. FINAL FOUR: This global map shows the top four contenders for MSL's landing site. The potential sites are boxed—Holden, Eberswalde, Mawrth Vallis, and Gale. *Image from NASA/JPL-Caltech.*

"It turned out that they [each site] received exactly the same number of votes!" Grotzinger chuckled as he recalled the conundrum. With almost one hundred scientists voting on four sites, the result had been a dead tie. What were the odds of that?

"It was unbelievable," he said. "Then I realized, wait a minute here. What we will do is to have people vote for secondary favorites as well. When we did that, there was a big difference. I realized something very important—it was clear that we are going to be divided if we picked one versus the other." However, if the resulting selection was at worst everyone's *second* choice, there would be few ruffled feathers. So they moved into the final phase of the process.

Grotzinger told them, "Pick your first favorite, pick your second favorite, pick your third favorite, and pick your fourth." He continued, "One site fell away right off the bat. The third looked a lot like the second. Then we decided the best thing to do was to have nine of the principal investigators pick the site." They were down to two candidates.

The results of this final phase were remarkable. "Gale was the fifty-percent favorite as everyone's first choice and eighty-percent favorite as their second choice."

The other primary choice besides Gale Crater had been Eberswalde Crater. It was agreed that this was a former river delta and that there would be good science to be done there. But "the problem is, now that you are there, what do you? What do you do for two years? What do you do for ten years? We characterized this site as a one-trick pony. If the trick works, it will be a beautiful day in the history of Mars exploration; but if it didn't work, then you will be faced with this awkward situation to try to explain to the taxpayers. They could say, 'We knew there was water on Mars before you landed, that's what MER showed. Now at the Eberswalde delta you [had] more water, you [had] a lake, but what good is your lake if it doesn't have organics in it?'"

What good indeed? They realized that at Gale Crater they would have far more options. There was the alluvial fan at the projected landing site and lots of interesting areas around it. Then, at the center of the crater, projecting 18,000 feet into the thin Martian air, was Aeolus Mons, also known as Mount Sharp. This huge feature was highly unusual in that it was composed of layer after layer of sedimentary strata. How it came to be in the center of Gale is still a topic of some conversation, but the important thing was that it was there. The mountain could provide a look at a huge chunk of the geological history of Mars, billions of years of fossilized data. Driving through the foothills of Mount Sharp would be the frosting on the cake.

"When you look at Gale, there's something for everybody. There [are] all kinds of different options. The risk was [that] there wasn't just one beacon gleaming with a definitive story there. Before we landed, people said, 'Yeah we know you guys see that alluvial fan, but it's kind of small. It's not the best-looking alluvial fan. Maybe we are going to get burned . . . maybe it's a lava flow.'"

While this sounds like a geological beauty contest, the point is that from orbit one thing can mimic another. He continued: "Lavas can spread out like cake batter. Every other site where we ever landed, we've been burned. Pathfinder was supposed to land in a field of what should have been rounded boulders. That team [later] published the interpretation that they were transported in water, but it was a hard sell. Then, Spirit lands on what is supposed to be a lake but turns out to be lava flows. Opportunity lands, and it's supposed to be on a volcanic edifice that's been oxidized and had its top crusted over with iron, but it's not. It's ancient, windblown sand, and it has goofy sulfate minerals that carry a lot of iron. Who the hell would have thought that?"

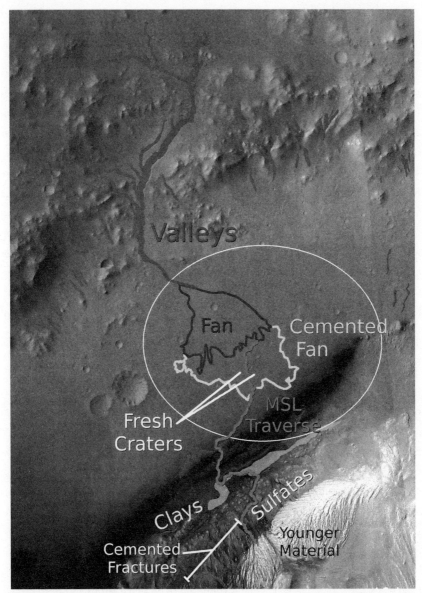

Fig. 9.2. GALE'S GRACES: This chart shows the many potential rewards of investigating Gale Crater. From the top, we see large, shallow valleys coming from the crater's rim, the alluvial fan that John Grotzinger often refers to. At the bottom lie the suspected clays and sulfates at the base of Mount Sharp that the science team covet, along with the billions of years of strata in the mountain itself. *Image from NASA/JPL-Caltech.*

Who indeed? Certainly not me, and apparently not a lot of much-smarter people, either.

A clarification is in order: these interviews were a lot of fun. Besides the interesting retelling of the early history of the program, and besides having the opportunity to chat with the people making the science happen, there are also moments like the above: "Who the hell would have thought that?" as Grotzinger sits, hands spread, asking what to him and the other scientists on the mission is probably as obviously odd as a Doberman driving a BMW is to the rest of us. I nod knowingly during moments like this, trying to imply that I get it, that of course I would have had the same reaction to the obvious. No chance, of course, but it's fun to feel like an insider, even for just a few hours.

The overall message was, however, clear: Mars is never short on surprises, and it's not stinting in the creation of mysteries. What unexpected delights—or agonies—would Gale Crater hold for Curiosity?

CHAPTER 10

THE SCIENCE PLATFORM

Curiosity is a mission with a purpose. Explaining that purpose to the public has been a bit of a challenge, however. With Pathfinder, it was easy: we wanted to send an inexpensive, quickly designed, experimental rover to Mars to drive around and look at some rocks. It's going to be a test-bed mission. MER was a follow-up to that first rover with a spectacular twin-rover mission. They would be bigger and more capable and would drive as far as they could. Their instrumentation allowed them to better see the landscape, drive through it with some autonomy, approach rocks, and examine them in greater detail. And, as always, continue the search for a wet, watery past.

Now, here comes MSL, and it has a vastly greater reach in scientific capability. But that statement alone does not convey what a quantum leap the mission, and the Curiosity rover in particular, represent. To better understand this, a discussion of the instrumentation on board is in order.

Recall that this is by far the largest and heaviest American robotic explorer to land on another world. The dimensions of the overall MSL spacecraft, in its flight configuration, is often compared to the Apollo lunar-landing program: it is fifteen feet in width, and its heat shield is larger than Apollo's. As for the rover, it weighs four times as much as the Apollo lunar-roving vehicle. The other popular comparison is that it is roughly equivalent in size to a Mini Cooper. And so forth. In short, it's the biggest, baddest boy yet on Mars.

All these comparisons vanish when you approach Curiosity in person—at that point it is just a huge, complex, and impressive machine. There are three versions that civilians like myself can see at JPL: One that is very close to the real item resides in the lobby of one of the main buildings on campus. Another two live up in the Mars Yard, the test track JPL uses to work out the kinks in an environment as close to Mars as they can get on-site. One of those, the Scarecrow, is the stripped-down, lightweight version that is tested in places like Death Valley. But stand next to any of them, and one thing pops out at you immediately: Curiosity is *big*. The top deck reaches to the middle of my chest. The camera mast towers over my head.

And then there are the wheels. Each of the six is the size of a small beer keg, both in diameter and in width. Of course, they weigh a tiny fraction of a keg—they are exquisitely machined to incredible tolerances.

Everything about Curiosity dwarfs previous rovers.

Its scientific capabilities are a match for its physical size. Not since the Vikings has anything this well equipped landed on Mars, and it blows that 1970s vintage machine into the weeds with its sophistication.

It's hard to encompass all the capabilities of MSL—the topic could fill another book. But let's take a look at what's under the hood. There are ten instruments, with a total mass of about 165 pounds, or fifteen times that of the MER rovers. What follows is an introductory inventory so we all know what's coming.

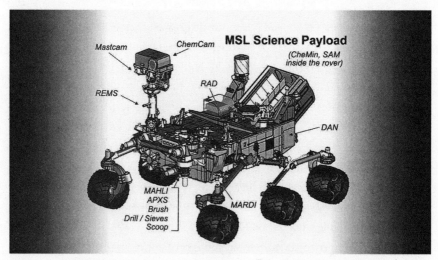

Fig. 10.1. CURIOSITY'S LOAD: The one-ton rover is packed with instrumentation, as indicated on this chart: the DAN and RAD instruments are located on top and on the side of the rover; ChemCam, Mastcam, and REMS are on the mast; CheMin and SAM are inside the main chassis; and the MAHLI, APXS, rock brushes, drill, and scoop are all stationed on the robotic arm. *Image from NASA/JPL-Caltech.*

So, on board Curiosity are the following:

CAMERAS

Most apparent is the camera mast. Both MER rovers had them, but this one is, of course, larger and more capable than what has gone before. There are two sets of traditional cameras on the mast, one with wide-angle and the other with telephoto lens systems. Early on, a zoom lens was proposed and development was begun, but it was just too difficult to pull off with the kind of reliability that was needed, so they settled for a twin fixed-lens system. The resolution of these cameras is stunning, making previous devices look like early cell-phone cameras. Both cameras have electronic memory built in, in the form of an eight-gigabyte buffer. This allows them to store over 5,500 raw images. The cameras can shoot more images at higher resolution and more quickly than their predecessors. They can also shoot 720p high-definition video footage at ten frames per second. It's not exactly cinematic speeds (a minimum of twenty-four frames per second), but it's far beyond previous efforts. Remember also that reliability is generally preferred over bells and whistles, so the fact that it has any HD-video capability is a huge accomplishment. Finally, these articulated on the mast to allow for full 360-degree pans.

There are also a number of lesser cameras on board: a total of twelve Navcams (for navigation) and Hazcams (for hazard avoidance) are mounted at all four corners of the rover. While they are generally used in far lower resolutions than the Mastcam, they have HD-capable chips inside as well.

Then there is the MAHLI instrument—the Mars Hand Lens Imager. It's called a "hand lens" imager because it is distantly related to the little magnifiers that geologists use on Earth to look more closely at rocks in the field. The MAHLI is mounted out on the robotic arm and can be placed close to anything that device can reach. At 1,600 by 1,200 pixels, it too is an HD imager. Next to the lens are some LEDs (light-emitting diodes) to illuminate rocks that are shaded or to operate at night. One set of LEDs provides regular white light and another provides UV light, which makes some minerals fluoresce. The mechanical focus mechanism can go from millimeters (almost touching the rock) to infinity, so it can also image the horizon if needed. Calibration targets are mounted nearby to check focus and color balance, including a 1909 Lincoln penny that is used for critical focus checks (the vintage penny was a favorite of one of the developing scientists).

The MARDI, Mars Descent Imager, is mounted to the bottom front of the rover. It had a short life—it was used only while the spacecraft was descending toward the Martian surface. The images it acquired—MARDI also utilizes a 1,600 by 1,200 pixel imager—would show the motions of Curiosity as it descended from the upper atmosphere, as well as give a progressively closer view of the landing site and its terrain as the rover neared touchdown. The camera was turned on shortly before the heat shield was popped off. It should be noted that funds for this device were "descoped," or trimmed, by NASA during the development process. But Mike Malin, the owner of the company that builds many of the Mars cameras, whom we met in Death Valley, contributed his own finds to finish the project. That takes the term *commitment* to a whole new level.

Fig. 10.2. EYES OF CURIOSITY: A self-portrait of the top of the camera mast taken by the MAHLI camera on the robotic arm. The large optic in the top box is ChemCam, the smaller, square lens sets below are Malin's Mastcams. *Image from NASA/JPL-Caltech.*

The final camera is called ChemCam. This is easily the most exotic of the cameras, at least in terms of how it functions. They are all amazing devices, but it is how this one goes about taking pictures that astounds. It is also mounted on the camera mast, but in the housing beside the camera is a powerful laser. In a twist on science fiction, NASA took a death ray to Mars. The device fires laser blasts powerful enough to vaporize rock, in packets of fifty or so nanosecond bursts, burning a small trail of holes in the target rock each time it is used. The telescopic camera next door sees the resultant light flash and converts that to spectral readouts, which tells scientists what the rock is composed of. Its accuracy is impressive, and it can be used from a distance of up to about twenty-five feet. The *Chem* in its name refers to the fact that it can discern the chemical constituents of the rocks it vaporizes. ChemCam was codeveloped by the United States and France.

Incidentally, also on the mast is the Rover Environmental Monitoring Station, or REMS. It measures various aspects of Martian weather, including humidity, temperature, air pressure, wind speed, and ultraviolet radiation. This was the only component damaged during the landing, when one of two wind-speed detectors was smacked by a small rock upon touchdown. But it still provides valuable data with a bit more effort on the part of the ground controllers. The instrument was provided by Spain and Finland.

ROBOTIC ARM

The next most obvious component is the large robotic arm below the camera mast and on the front of the rover. It is far larger than previous rover arms (no surprise here), and this size increase was required to accommodate the multitude of instrumentation it would carry. At the end of the arm is a large turret, about the size of the top of a barstool, that rotates the various instruments into position for use. The arm can extend up to seven feet, and the turret alone weighs 75 pounds— three times as much as the entire Sojourner rover from 1997.

The turret sports a variety of tools, including the aforementioned MAHLI instrument. In addition to this are:

The Alpha Particle X-ray Spectrometer (APXS): This instrument, contributed to the mission by the Canadian Space Agency (the same folks who brought you the Canadarm on the space shuttle and International Space Station), has flown in one form or another on every NASA Mars rover to date. When held near a rock, it bombards it with high-energy particles, then reads the signal that results from that irradiation. From this signal, the spectrometer can ascertain what elements are in the target. Among other things, it can tell whether or not the rock was ever exposed to weathering, water, and the like.

Scoop and Brush: For gathering rock and soil samples, there is a scoop on the turret. Nearby is a wire brush called the Dust Removal Tool or DRT (*dirt*, get it?), that can clean a rock prior to investigation by other instruments.

Drill: Perhaps the most remarkable part of the turret's instrumentation in terms of never having been done before is the drill. It's called the Powder Acquisition Drill System or PADS, and it works in concert with the Collection and Handling for Interior Martian Rock Analysis or CHIMRA. The entire combo is fancy nomenclature for a percussion drill that beats and grinds rocks into powder, then collects those grains, transports them into a couple of small chambers, and retains them for deposition into the body of the rover where the analysis stations are. Each of these samples is about the volume of a baby aspirin. If it sounds simple, it's not. The development and perfection of the drill and collection mechanism was a huge endeavor on its own, and the engineers spent a lot of sleepless nights working on it right up until launch. More on the drill and collection mechanism later—it's a fascinating story all on its own.

Now for the instruments on board the rover's main body. Hold on, because this gets complex.

We'll start with the easy ones first.

DAN

The Dynamic Albedo of Neutrons instrument: This device, provided by the Russian Space Agency, sends neutrons into the soil beneath Curiosity and then measures the return signal. Using this method, water content as low as a tenth of a percent can be measured as far as six feet below the surface.

RAD

The Radiation Assessment Detector instrument: This passive radiation detector measures radiation, that's it. It was turned on during MSL's cruise phase between Earth and Mars to measure radiation in the deep-space environment, and it continues to function on the surface of the red planet. Its primary purpose is to ascertain whether human voyagers to Mars would be able to survive the radiation levels on the way to the red planet and while on the surface. So far the answer seems to be yes, but with conditions. They will need a lot of shielding of one kind or another, both during the trip and while on Mars.

Now we get to the really juicy stuff:

SAM

The Sample Analysis at Mars instrument. SAM is one of two scientific superheroes of this mission. This machine takes up almost half the science payload of Curiosity, and with good reason—it's the costar, along with CheMin, of the show. It contains chemical and analytical equipment that not long ago would have filled a moderately large university laboratory, and it helps MSL to deserve the "laboratory" part of its name. SAM's primary chore is to find various forms of carbon and lighter elements associated with life, such as hydrogen, oxygen, and nitrogen. It does so with three subinstruments. Inside it contains a mass spectrometer, which separates elements and compounds by their mass for measurement and identification. SAM also houses a gas chromatograph, which heats rocks and soils until they turn into vapor, then analyzes the gasses as they escape. Finally, it has a tunable laser spectrometer. This incredible machine can measure the isotopes, or atomic numbers, of specific elements, such as carbon, hydrogen, and oxygen. All of SAM's instrumentation is highly accurate and has collectively raised the bar for Martian exploration.

CHEMIN

The Chemistry and Mineralogy X-Ray Diffraction instrument: Another remarkable powerhouse of analysis, CheMin looks at soil or drilled rock samples to determine types and the amounts of minerals within. The abundance of certain minerals helps to understand the environment of ancient Mars—for example, olivine persists where there is little water, hematite forms in wet environments. These are just two examples; many other minerals can be examined.

What is so ingenious about CheMin is the way it analyzes these minerals. Once the fine, powdered material is delivered to CheMin by the robotic arm, it is funneled into a small glass container. An x-ray beam is shot through the sample, and how that beam interacts with the sample is very instructive. Some of the high-energy waves are absorbed and fluoresce in a wavelength that tells the scientists what substance has intercepted it. The x-rays also react with the crystalline structure in the sample and become diffracted, and the angle and amount of this diffraction can identify elements within.

MEDLI

Finally, bringing up the rear, is a largely unsung device among the much-heralded instrument package, the MSL Entry, Descent, and Landing Instrument. Not to be confused with the MARDI camera, this device also had a short life, functioning well before MARDI did—it was active only during the entry and early landing phase of the mission, measuring pressures and temperatures and also allowing those on the ground to infer the exact direction of flight. MEDLI was mounted on MLS's heat shield.

This remarkable suite of investigative instruments gives Curiosity the analytical power of, quite literally, a science laboratory on Mars. That's one reason the name of the mission was changed from Mars Smart Lander to Mars Science Laboratory—same initials, but a vastly different meaning. The capabilities of a large and well-equipped analytical laboratory have been shrunk to the size of a coffin, placed in a chassis, and put on wheels.

So in general terms, the investigative sequence proceeds like this:

- Use the Mastcam to find interesting-looking areas or objects to investigate;
- Observe the target from a distance with the Mastcam telephoto imager;
- (Optional) Shoot a target with ChemCam's laser and observe the spectra of the burning rock;
- Drive to a target if it's worthy of investigation;
- (Optional) Use the MAHLI (microscopic imager) to look at the visual structure of the target;
- Use the APXS to analyze the sample in place;
- Scoop or drill the sample to get a bit of powder;
- Send the sample to CheMin, SAM, or both for analysis.

The primary purpose of the mission is to search for environments on Mars that were once habitable, that is, that could potentially have supported living things. There are many forms of habitability, but as currently understood, it will involve water and will probably date back to a denser, ancient atmosphere with more oxygen and nitrogen in it. These scientific instruments work in a beautiful partnership to find the answers the mission seeks. Once you are able to understand how they work together, it's a pretty incredible ensemble.

Even with the copious amounts of data provided by the sophisticated instruments, there is still plenty of abstract thinking for the scientists to do. The ultimate answers lie not just in the observations of the instruments but also in the context in which the samples are found. The result is a geologist's dream—the ultimate characterization of a landscape in the search for the biggest prize of all, life on Mars. NASA is very clear—Curiosity is not a life-science mission, at least not in the sense that Viking was. It seeks evidence of habitable environments and organic compounds. Of course, if along the way it happens to come across a fossil or two, nobody will complain.

CHAPTER 11

ROCKS, ROCKS, ROCKS!

If one did not know better, one might be excused of thinking that the exploration of Mars is only about looking at rocks . . . because largely, it is. It is also one reason I tortured myself with the complete span of geology classes a foolish undergrad takes until he encounters a semester of differential equations . . . but that's another story. The upshot is that, for some of us, studying rocks is fun until it becomes too much work. But to the JPL folk, the end of my struggle was their point of departure, and what they know about rocks and Mars is enough to make your head hurt. Fortunately, you have me here, writing this book, to spare you from at least some of it.

The study of Mars by machines, working on its surface, is generally focused on looking for past and present evidence of water. Where there is (or was) water there can be (or could have been) life. The past composition of the atmosphere is critical, as are moisture levels, and indications of atmospheric pressure, temperature, and many other things. Also, what elements were present? NASA uses a widely accepted acronym for these primary elements of life—CHNOPS, which stands for carbon, hydrogen, nitrogen, oxygen, phosphorus, and sulfur. These chemical elements seem to be essential to life in forms that we understand.

What do these past conditions and chemicals have in common? Well, some can be retained in one form or another in the geological record, and others can be inferred from what is observed. A lot of insight can be gleaned by looking at rocks, and specifically the deposition of what kind of rocks and how they changed over time. There are many reasons John Grotzinger was selected to lead MSL, and primary among them is that he is a premiere sedimentologist. He is interested in—among other things—rock and silt layers deposited over time in watery environments—and once the Pathfinder and especially the MER rovers found recurring evidence of this, he became the right man for the job.

Let's start with a look at what this field encompasses, since it has turned out to be so important for Mars exploration.

Sedimentology is concerned with the study of sediments as created by water over time. A sedimentologist will observe processes taking place in the field today, then apply that knowledge to what is found in more-ancient layered deposits. In general terms, the deeper or lower the sedimentary layer, the older it will be. And these processes appear to be universal—that is, they hold true on Mars just as they do here on Earth. So, clearly, deposition of sand and silt via water—over time and in amounts both great and small—are important to an understanding of Mars, and especially to gain a picture of the possible habitability of an area.

To further refine:

> The history of Mars exploration can be characterized by a series of exciting discoveries that have dramatically overturned previously held beliefs about the planet. Until very recently, the dominantly held position within the scientific community was that while geologic and climatic conditions during Mars' distant past may have been conducive to the potential origin and evolution of life, conditions on Mars today offer slim hope for life as we know it due to the unlikely existence of near-surface liquid water environments. However, recent results from NASA's Phoenix Lander and Mars Reconnaissance Orbiter missions suggest that present-day Mars may in fact contain a range of potential liquid water environments.

This accurate, if wordy, explanation came from the promotional materials from the Mars Habitability Conference held at UCLA about six months after the landing of Curiosity. I was privileged to attend, and some of the smart folks we have met here—including Ashwin Vasavada—were present. Of course, more has been learned since the conference was held in February 2013, but the idea of habitability remains the same—when, if ever, could Mars have supported life, microbial or otherwise?

Habitability, then, is the driving force behind the mission of Mars Science Laboratory. This is also what specifies the difference between a mission such as MSL's and that of Viking in the 1970s. Where the latter had operated on the assumption—or hope—that there might be microbial life on the surface that could be measured via a series of relatively simple chemistry-based tests, MSL operates on a series of new assumptions and hypotheses, developed over the intervening decades by further observations of Mars and of Earth. The Martian data has been gathered both from orbit (Mars Reconnaissance Orbiter, among others) and the surface (Phoenix and also the MER rovers). Intensive observations and explorations of our own planet have also greatly expanded our understanding of these processes.

A few examples of Earth-based data, and how it might apply to Mars, are in order. As regards life, and what it might look like, one need go no farther than the life-forms found since the time of Viking on Earth's ocean floor. I previously mentioned geothermal vents—the "black smokers," hot, mineral-rich torrents of ultra-hot water that bubble up through the ocean floor. These vents are usually found at the boundaries of tectonic plates or other areas that are volcanically active. The life-forms found adjacent to geothermal vents range from bacteria to worms, and all can exist in conditions previously thought impossible—the water can get up over *800 °F*. So the water is *hot*, there's no light down there, and the creatures present ultimately derive their nourishment from the mineral content of the upwelling water. Remarkable as that sounds, they are just one form of what are now known as extremophiles, forms of life that defy previous expectations. I mentioned some found in places like Antarctica as well, but there are also extremophiles in the Atacama Desert (some of which is officially the "deadest" place on Earth), protected by thin layers of rock yet still able to draw energy from the weak sunlight that penetrates. The life-forms in the Atacama are especially interesting, since that region comes about as close to Mars on Earth as one can get. It's deadly dry, gets tons of UV exposure from the ruthless sun, and even has nasty perchlorate in the soil—just like Mars. Yet life manages to find a way, even there.

It's interesting that while NASA will not refer to MSL as a life-science mission or "life-detection mission," it happily calls it "the first astrobiology mission since Viking." The distinction is small but important, if unintentionally misleading. MSL is not seeking life (though mission scientists would be ecstatic to find some definitive sign of it), but potential habitability, past or present, and possible biosignatures from current or former living things. These would likely take the form of isotopes of organic carbon.

More specifically, as spelled out by Michael Meyer, MSL's lead man at NASA, during a public presentation in 2012, the goal of MSL and Curiosity is to "explore a region of Mars and determine if that area was ever able to support microbial life and assess its potential for the preservation of biosignatures." That last part is important to us because for the most part, any biosignatures existing will come from rocks and soil (there has been hope of finding a biological source of methane on Mars, but this has, to date, been unsuccessful).

Fig. 11.1. EXTREMOPHILES: This image of a "black smoker" geothermal vent on the ocean floor is an example of the new ways we have learned that life can adapt. The vents can emit dark, mineral-rich water heated to over 800°F. Below the plume are the various worms and other life-forms that live directly or indirectly off the vent's emissions. *Image from NOAA Okeanos Explorer Program, INDEX-SATAL 2010.*

This idea formed the backbone of a program strategy called "the Exobiology Strategy for Mars Exploration," which was defined in 1995. It underwent several revisions, many due to the work of the MER rovers Spirit and Opportunity. Even as MSL was being proposed, approved, developed, and built, this idea was in flux. And it will likely change again, based on the scientific results from Curiosity's explorations. Part of the process was to limit NASA's Mars exploration goals, to reign in overly general expectations and the expenses that they could incur. These goals were refined from the broad proclamation of *exploring overall Mars habitability, Mars polar habitability, and its recent climate*, to this simpler, more focused directive for MSL: *assess the habitability of Gale Crater, past and present*. It is a more-manageable set of expectations for a such a mission.

This mission, once it was pinned to Curiosity, was expected to last at least one Martian year, or about twenty-three months. JPL'ers have learned much since Pathfinder and the MER rovers, and they have refined and evolved not only the techniques for the design and construction of the new rover but also their expectations of (and confidence in) their machines.

As we know, once the landing-site selection process had concluded, Gale Crater was the target. Ninety-six miles in diameter, it is dominated by Mount Sharp at the center, 3.5 miles high—three times as high as the Grand Canyon is deep! Mount Sharp may look at first glance like central peaks found in lunar craters such as Tycho, but that is where the similarity ends. Tycho's peak, and most similar peaks within craters, are thought to be created during the impact that causes the crater to form. Shock and energy returning from the impact causes the central peak to rise. Mount Sharp is nothing like that.

Instead, it was laid down over the 3.5-billion-year life of Gale. It has sedimentary and cross-bedded layers that have been seen both from orbit and from Curiosity's long-range cameras. It appears that this enormous mound formed when water flowed into the crater and, possibly aided by wind-blown deposits, filled it with sediment. As the water dried, it left layers behind, which built up over billions of years to approach the larger earthly mountains in height, and fill the crater. The remaining central location is due to wind patterns coming over the edge of the crater, scouring away material in an ever increasingly complex pattern of swirls. The net result: the perimeter material is gone, and a huge mound of it remains at the center.

Among the layers of Mount Sharp are clays and other mineral types indicative of water in the past. This, along with the other data flowing in from Curiosity, sealed the case for water in Gale, and lots of it, and over a very long period. Mars was, at one time, very, very wet. And water, of course, is a key part of the ancient habitable environment story. Mount Sharp, whether or not it contains evidence of organic materials, appears to tell that story across time, layer by ancient layer.

Of course, there is more to Gale Crater than just Mount Sharp. One of the reasons it was ultimately chosen as the MSL landing site was the rich variety of geological and topographic features within the crater, which offer many other opportunities for discovery.

The landing zone, ultimately named Bradbury Landing after the famed science-fiction author who died shortly before Curiosity arrived at Mars, is near the terminus of an alluvial fan. These features occur when water washes down a hillside over a long period of time, depositing rocks, sand, and silt at the base of the incline. It's a great place to study materials that have migrated from higher elevations, and some of the stuff observed there will have traveled many, many miles, both horizontally and vertically. It's like having rock and soil samples from the entire region collected in a relatively small area for you to inspect at your leisure. And that's just one feature of Gale.

There are also areas that incorporate a characteristic called "high thermal inertia." This simply refers to the way heat, absorbed by rocks or soil during the day, gradually radiates away during the night. Gravel, sand, and loose soil lose heat quickly, solid rock and hardened sediments do so much more slowly—they tend to retain heat. It's similar to the way a concrete house insulates better than a wooden one, and why sidewalks exposed to the sun stay hot well into the night after a long summer day, while nearby loose dirt cools rapidly. With infrared cameras that can image temperatures visually, the geologists have built up thermal maps of the region over time from orbit. By interpreting these maps, they can determine with some accuracy the probable structure of the terrain on the floor of Gale Crater, as the temperatures on Mars can change from day to night by a couple hundred degrees Fahrenheit or more.

Now there are nuances to this as well. Areas that are made up of what is called conglomerate, which is a kind of natural, sedimentary cement that binds gravels and clays and what have you, also behave like solid rock and hold heat longer. This thermal behavior, if seen in an area that appears to be comprised of water-borne deposits as opposed to solid rock, can be an indicator of something interesting to go and investigate.

It's all about the rocks, you see . . . and sand, gravel, clay, dirt, and more. They are the timekeepers, and living record, of Mars's history.

So, when you are seeking a "habitable environment," it is presumed that you need water (as would be found in regions like the one just described), carbon, a friendly atmosphere, and a form of energy to allow life to exist and possibly flourish. Indications of some of all of these may show up in layered rock formations.

Carbon compounds can be identified by the Curiosity, utilizing ChemCam, from a distance. This can aid in deciding what to drill or scoop for a sample. Once the sample is inside the rover's laboratory, SAM can determine the molecular weight and charge information, which can aid in identifying the carbon's likely origins—organic or inorganic.

Carbon can be a bit of a red herring on Mars, however. Even organic forms of carbon can be the result of a meteor fallen from space—the carbonaceous chondrite variety, as opposed to resulting from life processes. These meteorites, rife with water and organic compounds, are remnants of the formation of the solar system. They wander through interplanetary space for billions of years until they encounter something to stop them—in this case, the gravitational field of Mars. While the carbonaceous chondrite variety comprise just a small percentage of all meteors, they fall to the surface of that planet all the time, so it follows that much of the carbon present there will be meteoritic in origin.

One more thing about organics and rocks—unique circumstances must occur for organics to be preserved and observable in sediments. Mars takes the full brunt of the sun's radiation—including ultraviolet and cosmic radiation. It has a thin atmosphere as well as a negligible magnetic field, which stops little of either (Earth has a dense atmosphere and a strong magnetic field, both of which protect us from the worst of the sun's intense energy). Martian soil also contains perchlorate, a chemical that is not helpful to detecting organics and that is toxic and corrosive to potential life-forms or their remains. And while perchlorate can serve as food for some microbes observed on Earth, on Mars the presence of perchlorate in a sample that it is heated (as most experiments on Martian soil have done) can destroy organic compounds.

Mars hides its secrets well. Just ask the guys looking at the rocks.

Clearly, understanding the story of sedimentation on Mars is a vitally important part of unraveling its history. But not that long ago, planetary scientists weren't even sure if there were *any* sedimentary deposits on Mars. The Vikings were static landers that remained right where they set down, using robotic arms that could reach out only in a ten-foot-or-so arc with their scoops. Pathfinder was limited to a thirty-foot radius from the lander. But the fleet of orbiters that have been staring at Mars since Mariner 9, culminating with the Mars Reconnaissance Orbiter, offered spectacular visual resolution that showed a lot more detail than ever before. What had to be inferred earlier could now be seen directly. But even then the jury was out—is what we saw below the result of volcanic activity, as most believed, with some wind and water mixed in, or had major sedimentary events that had taken place?

Mike Malin, whom we have met briefly as the creator of the cameras for Curiosity and other Mars machines, tipped the balance in this debate. Malin has a deep passion for Mars and, along with his close associate geologist Ken Edgett (whom we will get to know better soon), has devoted untold hours looking at images of Mars, mostly from his own cameras. In fact, hundreds of thousands of orbital images of Mars have been returned from his cameras alone. Malin could be likened in enthusiasm to Percival Lowell, the late Victorian who devoted thousands of hours to observing Mars with his telescope—except that Malin neither allowed his imagination to run away from him, nor did he map his own eye's retinas as Lowell may have. But Malin did make an intuitive leap, as recounted by Grotzinger. The paper that Malin and Edgett authored about their observations caught Grotzinger's attention because it contained a theory ripe for a sedimentologist like him to latch onto and start looking deeper.

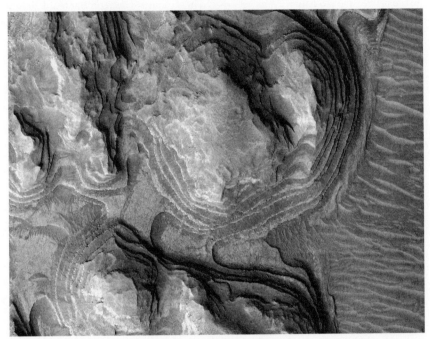

Fig. 11.2. SEDIMENTATION SQUARED: This Mars Reconnaissance Orbiter image shows an area on Mars called Arabia Terra. This is an extreme example of weathered sediment—few are as obvious as this. But the extreme nature of the terrain does show how sedimentary layering can be seen from orbit once one understands what one is looking at. *Image from NASA/JPL-Caltech.*

Grotzinger started the conversation that day discussing "hazing" in science, which I thought was a great term for the way scientists question each other. More specifically, he was discussing how, when there is an established and largely accepted body of theory, there is pressure to "get on the bandwagon" and just go along with prevailing theory, unless there is clear and compelling evidence to the contrary. That evidence is what Malin and Edgett had labored so many hours to find, and Grotzinger recounted their quest.

"Those guys spent a lot of time looking at Mars; they really were looking incredibly carefully in an attempt to make a distinction between things that were obviously volcanic terrains associated with obviously volcanic features, as opposed to these giant Noachian [an ancient geologic period on Mars ending about 3.6 billion years ago] craters that were filled up with something that looked layered when there was no other crater nearby." These features looked like evidence of sedimentation . . . but were they? If they were, this would indicate a lot of water doing a lot of work at one time or another in Mars's history, and probably very long ago. Following this line of thought, Eberswalde Crater caught Malin's attention.

"In 2003, the Eberswalde Crater paper was published by Malin and Edgett, [and] that was a barn burner," Grotzinger recalled. "But the way that science works is that when there's a reigning paradigm, people don't spend a lot of time questioning it, they would rather run with it." Malin and Edgett were swimming upstream, and the current was running against them in the form of contrary opinion. But Malin was used to going against the prevailing current—decades before, when he first suggested to NASA that high-resolution cameras orbiting Mars could return valuable data, he was told that they already had all the images they would ever need from the Viking orbiters. Well, as you can guess, whoever at NASA said that was wrong. Viking was like a cheap pocket camera compared to the high-powered optics that Malin had in mind.

But back to hazing in science: "We all do it. We all do it to each other. It's an inherent fallibility of science. Every time there is a band wagon, you will always see somebody saying 'Why can't this person get on board?' But one time in ten, that person is at the leading [edge] of [a] revolution, and that is where Malin and Edgett were. They said 'These features are inconsistent with being just lava flows; there is more here than meets the eye,'" Grotzinger said. And the more people looked, the more it seemed that what they saw in some areas could be the result of millions of years of sedimentation, and those sediments could yield a goldmine of information about how much of Mars was formed, where the water had been, and what it had been doing when it was there.

Another revelation resulted from Malin's interpretations: it appeared that not all watery activity had been ancient. Some of these changes looked relatively recent to his eye. The idea of water existing as a liquid on modern Mars was a revelation (and in some quarters, heresy). Besides being a shock, given the incredibly low atmospheric pressure, it also implied possible supplies of water that could be utilized by organisms today. Note that nobody is waving their arms and proclaiming, "Hey! We found a place where there should be life!" It's simply one more link—albeit a more current one—in the chain of events that could contribute to a habitable Mars.

The main point was this: the discovery of sedimentary rocks on Mars, among its other effects, was primary in shifting the search for life from the Viking model ("Are there critters here?") to the MSL model ("Was there an environment on Mars that could have, at one time, supported critters?").

Being able to drive a rover up to exposed sedimentary layers—"outcrops," in the parlance—would satisfy a lot of needs. First, it gives you a visual record of geologic time, especially in a place like Mount Sharp where the sedimentary layers reach very high (note, however, that this is all *relative* time, so you also need to look for signs of ancient events that might "fix" some part of the layer to a specific period).

Second, sedimentary layers tell the story of water through a combination of things frozen in time. There are the sizes of the objects in the layer—silt says one thing, sand another, pebbles and larger items still another. Generally they indicate the speed and force of the water. The chemistry of the layer tells you other things—how wet, how salty, and so forth.

And there is a third item, which for now is a fantasy but may not stay that way. The MAHLI instrument (microscopic imager) was not designed to look for micro-fossils, and in any event it would be sheer serendipity to stumble across some. But it could detect a huge collection of them—say a preserved colony of microscopic creatures—or a larger life-form, such as a snail. And where might the record of such a colony be found? In sedimentary layering—the preserved remains of an ancient seabed or the like. And this brings us back to an item in the opening chapter of this book—stromatolites.

Stromatolites are rock layers that have a shape that can—sometimes—identify ancient life-forms. They often indicate microbial colonies or mats, and on Earth they can date back as far as 3.5 billion years, about as far back as you can see evidence of *anything* biological. They come in many shapes and sizes and can be found all over the planet. But they can also be chimerical, which is to say that a stromatolite might indicate life—biological activity—or it might not.

This was made clear to me in a profoundly visual way at Caltech during one of my meetings with Grotzinger. It helped to illustrate why his undergraduate classes would be popular.

Fossils and rocks adorn much of Grotzinger's office. He showed me a number of stromatolites, ranging from flat slabs with knobby textures to more rectangular rocks with a cross-section that showed the same knobby layer from the edge. I said, "Oh, so this is what microbial life looks like when preserved as a fossil." We have all seen fossilized shells, plants, and dinosaur bones, but it was the first time I had knowingly held microbial fossils in my hand. They are distinguishable only because they grew into a colony large enough to see and preserve.

Of course it's not that simple—that would be too easy.

He handed me another rock—a rectangular, white one about five or six inches long, three inches wide, and three inches high. The cross-section showed more of the stromatolites in profile. It looked just like the other stromatolites.

I looked at him and waited for the punch line. "That's not a result of biology." He said.

It turns out that what I was holding in my hand was a chunk of brake lining from a railcar. When the stony material used in this lining gets hot enough (as it does through the friction of slowing a train), the heating creates wavy lines in the "rock" as chemicals separate, change, and create visible divisions. And they look a whole damn lot like the *other* stromatolites, the ones that I had just held and that resulted from biological creatures. It was an enlightening moment.

This is an extreme example. There are lots of other nonbiological activities that can cause rocks to look like the result of living processes. On Earth it's often extremely difficult to discern whether or not the rock you are looking at is a result of biology or nonbiological activities in the ancient past. This differentiation can be a challenge on Earth; you can imagine how difficult it would be on Mars.

So in addition to designing the spacecraft, building it, launching it, landing it, and figuring out where to go once you are there; once you have located sedimentary layers, and even if you find stromatolites, you may *still* not know how they formed—critters or no critters? This is where geochemistry comes into play. *If* you can find the structures, and *if* you can find a biosignature within them, then you *may* have some indication of life on Mars.

Of course, the scientists will argue over that, too. It's what they do.

CHAPTER 12

EARTH ATTACKS!

The seminal Hollywood fabrication about biological pollution from outer space may be *The Invasion of the Body Snatchers*, a seldom-seen black-and-white pot-boiler about nasty creatures from space that invade Earth and exterminate people, growing exact (but nefarious) copies of them in giant seed pods. Today the film is seen as a by-product of Cold War hysteria that produced a number of such fantasies. A remake from the 1970s did little to improve on the basic idea. Alien species coming to Earth rarely bode well (Spielberg's *E.T.* excepted).

Many other movies, TV shows, and books have expanded on the invasion theme (often with human-sized, rubber-masked alien beings), but the point is this: you don't want evil things coming to your planet. When the time came to start sending spacecraft to the surface of Mars, it occurred to some within NASA that it might not be such a great idea to make the same mistake in reverse, either.

The issue of the contamination caused by space exploration was outlined as early as 1958 by the International Astronautical Federation, and in 1967 the United Nations issued a report with the weighty title "Treaty on Principles Governing the Activities of States in the Exploration and Use of Outer Space, Including the Moon and Other Bodies." Along with many other proclamations, it included wording about keeping other planetary bodies safe from earthly infestation, specifically that all countries signing the treaty "shall pursue studies of outer space, including the moon and other celestial bodies, and conduct exploration of them so as to avoid their harmful contamination." Exactly how this would be accomplished was left to other experts.

When the Viking program was being planned, contamination of Mars became a central concern. Letting earthly bacteria loose there to reproduce, mutate, and eventually devour future astronauts was not on anybody's hit parade. More important to that particular mission, the life-science people did not want to fly all the way to Mars, land, sample soil, put it into their billion-dollar spacecraft, and end up finding life, only to realize that it was actually something that hitched a ride from Earth. They had to figure out a way to sterilize the landers.

Their solution may have been a case of overkill, but if so, it is overkill that has continued to some degree to this day. There was—and continues to be—a lot of debate over how careful we need to be in this regard. Many scientists believe that the surface of Mars is so hostile, given the high radiation, extreme temperatures, and the nasty chemicals in the soil, that few, if any, earthly life-forms could persist for long once they arrived. But as your mother probably told you, one can never be too careful.

After the Vikings were built, they were subjected to various indignities to sterilize them as well as 1970s science knew how, just short of destroying their innards. In fact, about ten percent of the mission's budget, $100 million, was diverted to this endeavor. They called it Planetary Quarantine.

First, the machines were built in a clean-room environment—white gloves, breathing masks, little hats to cover the hair, bunny suits, and booties. The room's air was processed to make it as sterile as possible. But this can never catch everything.

The components and final assembly were cleaned obsessively. And I don't mean the type of mop-down they do each night at the local McDonald's. I mean alcohol and tiny paintbrushes and sterilized Q-Tips in every nook and cranny of the spacecraft. But that's not all.

They were then sealed in giant ovens and baked at 257°F for thirty hours; five hours was considered sufficient to kill just about anything so, in a tribute to their thoroughness, they cooked the landers six times as long as needed. The goal was to reduce the number of bacteria, or spores, to no more than roughly 300 per square yard, or about 300,000 total spores per spacecraft. This remains the goal for planetary protection even today. Note that we are talking about spores here, things that can survive harsh conditions and potentially reproduce. Active microorganisms are thought to be insufficiently hardy to survive the Martian environment (or the harsh transit from Earth), but nobody was taking any chances.

The landers were then sealed up and not touched again prior to launch. In all, during the cleaning and sterilization processes, six thousand tests were performed to ensure that the probes would be as clean as science could make them. As far as anyone knows to this day, nothing survived the trip.

This may seem like vast overkill, and some in the scientific community would agree with that sentiment. It's tough on the machines, it's expensive to do, and it limits what can be built and flown. But it's worth mentioning that in 1969 when Apollo 12 landed on the moon, part of its mission was to retrieve parts of a robotic lander that had set down nearby three years earlier. The moonwalkers snipped off parts of the lander's camera, bagged them, and brought them home. Upon examination, biologists were surprised to find bits of *streptococci* virus still living on the pieces. Once subjected to the harsh vacuum of space and the unrelenting radiation of the lunar surface, they had merely taken a nap for three years until they came back to Earth. So the concerns about contaminating Mars (or other destinations) are not entirely misplaced.

In a slight diversion, let's consider NASA's Galileo spacecraft. It was sent off to explore Jupiter and its moons in 1989, arriving in 1995. The bus-sized spacecraft was assembled, like all of JPL's machines, in clean-room conditions, but it had never been subjected to the rigors of sterilization since it was designed to operate only in space. As part of its mission, it took many looks at Jupiter's moon Europa, and over time it became clear that there is probably a vast ocean of liquid water below the surface of the barren, icy moon. Where there is liquid water there could be life. When the mission wound down, NASA was concerned about unintended contamination of the potentially fertile moon by the massive spacecraft if it happened to crash on Europa. This contributed to the decision to send it slamming into Jupiter's dense atmosphere, where there was deemed to be little to no risk of meaningful contamination. *You can never be too careful.*

Then in the 1990s came the Pathfinder mission. Though it was operating on a lean budget, it was going to land on Mars and hence needed to be sterile. It was subjected to extreme cleaning similar to the Viking protocols but was not baked. There are many issues with baking, but the primary constraint is in the materials used on the spacecraft. Pathfinder was to be a highly cost-effective mission, so they used as many off-the-shelf components as possible. In general, these are far more susceptible to damage than items specifically created for spaceflight or military uses. Think of it in terms of setting your oven to 250°F and leaving your cell phone in it for a few hours. The result would be a thoroughly baked, and likely inoperable, cell phone.

Once Pathfinder arrived on Mars, the Sojourner rover sat atop the lander for two days. There were checks to be performed, but this was also a chance to allow the severe radiation that bombards Mars—including ultraviolet light—to kill anything remaining on the rover, especially the wheels, since they would come in direct contact with the soil. Later tests determined that even brief exposure to such conditions would exterminate about 90 percent of anything that might have hitched a ride.

When the Mars Exploration Rovers were being prepared for flight, similar wipe-downs were performed to kill anything that could be reached with alcohol-soaked swabs and sprays. Additionally, they used a method called "heat shock," baking it for fifteen minutes at almost 180°F. Again, the 300,000-spores benchmark was used. As Laura Newlin, JPL's planetary-protection lead for MER later said in a May 2003 press release, "Keeping the spacecraft as clean as possible before, during and after launch is very important for any science instruments searching for organic compounds on the surface of other planets. Up to 300,000 spores are allowed on the exposed surfaces of the landed spacecraft; that many spores would fit on the head of a large pin."

Then, along came the MSL project. While not a true life-science mission, the Curiosity rover would be carrying instruments of unprecedented sensitivity, so sterile conditions were critical.

The usual clean-room conditions were utilized. Frequent cleaning and wipe-downs were performed on all exposed surfaces. Components that could tolerate baking, such as parachutes, thermal blanketing, and other parts were baked at up to 230°F.

But in an innovative solution to the cleanliness of delicate parts, the rover's interior—which contains incredibly sensitive and somewhat-delicate components—was sealed. Venting of the main body of the rover, to equalize pressures, was allowed only through highly efficient and dense filters to keep anything living inside.

Receiving particular scrutiny were the items that would come in direct contact with Mars. As noted, the parachute and other items were baked to sterilization. But the wheels were a concern, and the drill bit was also a huge worry—it would be providing sample material to the instruments on board, and besides apprehension about contaminating the planet, they wanted to avoid false readings caused by non-Martian organisms.

But before launch there was a lapse in the protocols. It was with regard to the drill bits, a critical part of the sampling system. The design had called for the bits to be sterilized and sealed, and not to be touched by anyone or anything until after the rover landed and needed them. But the robotic arm held the drill head, which would not be able to function without a drill bit. The engineers worried that the drill's ability to grab a bit might somehow be impaired, so why not simply place one in the drill before launch? So, during the final phases of preparation, MSL technicians opened the sterile box and attached one of the bits to the drill mechanism. This way, if there was a failure to grab a bit, at least *one* would be available. It made eminent sense . . . unless you are the guardian of the cosmos regarding spacecraft sterility.

And there is such a person. It's currently Catherine "Cassie" Conley, NASA's Planetary Protection Officer. Shortly before launch, and well after anything mean-ingful could be done, she discovered the lapse of sterile procedures.

She continued the tale in a NASA statement: "As the Planetary Protection Officer for NASA, I am responsible for ensuring that the United States complies with Article 9 of the Outer Space Treaty . . . which specifies that planetary explora-tion should be carried out in a manner so as to avoid contamination of the bodies we are exploring throughout the solar system, and also to avoid any adverse effects to Earth if materials are brought back from outer space."

She repeated the challenge of interplanetary sterility: "For Mars, 'clean' in terms of spacecraft surfaces, regarding biological contamination, is that there should be fewer than 300 heat-resistant bacterial spores per square meter of spacecraft surface. There's an additional requirement for internal bacterial spores, inside a circuit board or inside the glue that's been used to attach two things together. But for surfaces, which are what you worry about for spacecraft that haven't crashed, it should be fewer than 300 per square meter, if you're going to a place on Mars that isn't given special protection. If you want to go to a place where there might be liquid water on Mars—a 'special region,' as it's called—it should be reduced by an additional 4 orders of magnitude, by some kind of treatment like baking in a dry-heat oven."

The reason the MER rovers had not been baked was that they were not expected to travel into "special regions"—no Martian beaches, lakes, or bayous. "It's extremely difficult to build a spacecraft that can tolerate several days of baking," she continued. "The cleaning procedure is something that a lot more materials can withstand than the baking, so in order to allow missions to go to Mars without such a stressful treatment, it was decided in the early 90s that, depending on what you were doing, it was okay to just do the Viking pre-sterilization levels."

When managers for MSL changed the process regarding the drill (that is, opening the covered and sterilized spacecraft and moving the drill bit from the rack to the drill), the memo somehow took a slow route to Conley's desk. It arrived "very late in the game," she said. So, while the drill placement was performed within a very clean environment, it was not up to NASA's planetary-protection specs. It was certainly an inadvertent oversight, but an oversight nevertheless. "That's where the miscommunication happened," said Conley. "I will certainly expect to have a lessons-learned report that will indicate how future projects will not have this same process issue. I'm sure that the Mars exploration program doesn't want to have a similar process issue in the future. We need to make sure we do it right." Hmmm. Point made.

Given the scope of the MSL mission and the complexity of the rover, this seems a small breach. But when you are charged with one sweeping mission—keeping other planets safe from earthly contaminants—you need to worry about these things. Upon further review, the issue turned out to be a small one. As far as they could see from orbit, Gale Crater did not have any potentially life-harboring ice within ten feet of the surface, the closest allowable for this mission's level of sterility. "That reinforced the reasonableness of not having the drill bits sterilized, because there's unlikely to be 'special regions' in the Gale Crater landing site," Conley said. And while the process was not exactly within spec, she noted that it was the cleanest lander since Viking—which is pretty damn clean.

There are issues beyond MSL. The next Mars rover, the Mars 2020 mission, may include the capability to "cache" samples, that is, to take the best bits of Mars they can find and situate them for later pickup and transport to a lander. A subsequent lander would gather them, and then rocket them back to Earth—a sample-return mission, currently the fondest hope of the robotic Mars exploration community.

But planetary contamination can go both ways. What if some nasty, human-melting, zombie-creating megabug came back hiding in the Martian rocks and dirt? In fact, this concern was first dealt with during the Apollo lunar landings. As far-fetched as it seemed at the time, there was concern that some . . . *thing* . . . from the moon might escape from NASA's labs and devastate life on Earth. So when the early moonwalkers splashed down in the ocean after returning from their mission, they were forced to don biological-containment suits—basically the opposite of what biohazard workers wear—to contain any such nasties their clothes or bodies might be carrying. The moon rocks were in sealed metal boxes and then whisked off to be placed in equally sealed laboratory quarantine. Once the astronauts got back to Houston, they spent three weeks inside a customized, sealed travel trailer. Blood samples were taken and tests were run. The air exhausted from the trailer was filtered and examined. Nothing was found.

Fig. 12.1. LOCKUP: The Apollo 11 astronauts can be seen in the window of a specially prepared trailer as they speak to President Nixon and wait out their quarantine after returning from the moon. They were kept isolated for twenty-one days, as were the crews of Apollo 12 and 14. After the latter mission, NASA decided that the risk was minimal and the practice was discontinued. *Image from NASA.*

It was only later that a journalist pointed out that when the astronauts, bobbing in the ocean after splashdown, opened the hatch to allow the navy divers to toss in the biological-containment suits, a *thing* could have just strolled out right then and entered the air or the ocean. Oh well, too late. Apparently nothing came back to Earth with them.

But what of Mars? In 1997, the US National Research Council released a study that concluded that the risk to Earth from the kind of extraterrestrial organisms we are likely to encounter on Mars is minimal—a view still held by most. "Contamination of Earth by putative Mars microorganisms is unlikely to pose a risk of significant harmful effects." But they added, "The risk is not zero, however."

This was updated by the same group in 2009: "The potential for large-scale pathogenic effects arising from the release of small quantities of pristine Mars samples is still regarded as being very low . . . extreme environments on Earth have not yet yielded any examples of life-forms that are pathogenic to humans."

So, in brief, we are unlikely to be transformed into green slime by Martian microbes. And it's a good thing, too, for millions of tons of Martian stuff has already reached Earth. Since the solar system formed, bits of Mars have been knocked off by asteroid collisions and meteoritic impacts, and have made their way to Earth, entering the atmosphere as meteors, with some hitting the planet. It is estimated that at least 10 percent of that total mass is not sterilized by heating. You may recall the famous Alan Hills Martian meteorite of 1994, which was found in Antarctica. Some researchers suspected the Martian rock of containing tiny fossils of living microorganisms (most have since concluded that the microfossils were probably not of biological origin, though the debate continues). This rock was just one of many that have been identified as being of Martian origin, and countless more have fallen over the eons—on Antarctica, on every other continent, and into the oceans. If there are Martian life-forms on Earth, we don't know about them.

As an aside, there is a hypothesis that dates back at least to 1903 called "panspermia," which theorizes that all life on Earth may have originated on Mars or some other celestial body. There are good reasons to think that this may be true, but no strong evidence has yet been identified.

Later in her NASA interview, Cassie Conley jokingly claims that she spends most of her time at NASA headquarters "answering e-mails," but her role is a very serious one with that dreadfully serous title. "Planetary Protection Officer" conjures up some compelling images. "It's always interesting to see people's expressions when I introduce myself as the Planetary Protection Officer. Most people think of the characters in *Men in Black* when they hear this title." She notes that when she took the job someone at NASA gave her a pair of Ray-Bans just like Will Smith wore in the movie. "That was pretty entertaining," she added.

Curiosity may indeed have transported germs or contaminants to Mars on its drill bit. If it did, chances are that the radiation hitting the surface of the planet probably did them in, and if not, the action of drilling in the highly toxic soil probably did. In short, chances are that it's harder to be a living microbe or spore on Mars than it is to be a punk-music fan at a Mozart concert. But only time will tell.

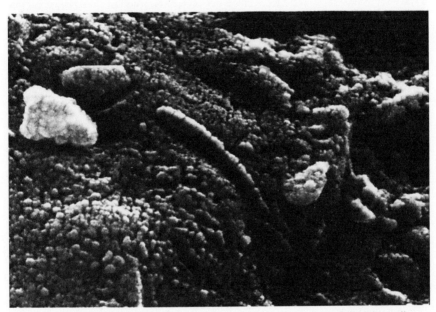

Fig. 12.2. DEATH FROM ABOVE? A high-magnification close-up of the Alan Hills Martian meteorite from Antarctica, ALH84001. While the central object does strongly resemble a microscopic worm and caused a sensation in the 1990s, the general conclusion was that it was far too small to have biological origins. *Image from NASA.*

CHAPTER 13

DELAYS, DELAYS . . .

No good NASA story is complete without a discussion of schedule slips and budget overruns, and MSL's tale is no exception. I do not mean to sound glib here—I merely wish to illustrate the enormous pressures, and frequent financial consequences, of operating in the complex, challenging, and ultimately flawed political and bureaucratic environment within which missions like MSL must proceed.

To begin with, when a mission on the scale of MSL is proposed, it is a bit like going back to 1961 when President Kennedy declared that the United States would land a man on the moon by the end of the decade. At that time, NASA officials were alternately thrilled and horrified (if they were smart, mostly the latter) by Kennedy's announcement. A manned moon mission was close to the heart of many in 1961, but on *that* schedule? With 1960s technology? At that time, the United States had exactly fifteen minutes of manned spaceflight experience (Alan Shepard's little suborbital hop in the Mercury spacecraft); we did not even know *what* we did not know. We didn't know how to build big rockets, how to keep people alive in space, if people *could* live in space, how to land on the moon, what metals to use, what fuels, and so on. Oh, there were ideas—a lot of them—but nothing tested or proven. And this kind of uncertainty is toxic to the engineering mind, at least when you are asking for a commitment to a firm delivery date less than a decade away.

MSL was, in many ways, similar. Of course, much was different—more was known about spaceflight, there would be no people aboard, it was Mars and not the moon, and so on. But there were many unknowns, including how to land this massive machine on a planet a couple hundred million miles away. The MER rovers had not even begun their mission when MSL was being conceived and was seeking official approval. How the heck do you create a cost figure and a delivery date for something with so many unknowns?

The temptation would be to bid high and build in a slush factor. To a degree, that's what Wernher von Braun did in the Apollo era, and consequently he looked pretty damn good when he delivered the goods mostly on time and on budget. But that was the space race and funds were flowing to NASA like a river. MSL was being proposed in the late 1990s and conditions were not even remotely similar. If JPL were to estimate as high as they would need to in order to ensure on-time delivery of a quality product, some members of Congress would balk. Actually, make that most. Just imagine trying to get a huge dollar number out of a representative who's district was not building any part of your spacecraft, who had no part of the launching or controlling of it, and who is much more concerned about protecting farm subsidies than exploring whatever the heck is on Mars. It would be a nonstarter; and most high-budget planetary-exploration programs have been thus since the end of the Viking days.

So the solution? Bid low, try your best, and then let fate take a hand. Of course, this means cost overruns. Hopefully they are within a reasonable range and will not cause too many officials to turn purple as things move along. And—hopefully—by the time people *do* get upset, you will be far enough along the road that the project will not get canceled.

With MSL, it turned out to be a close thing.

If you bring up this bit of discussion at a dinner party with some nice NASA folks, you might not get quite the same version that I am presenting. This is greatly simplified and there are a thousand and one nuances I am glossing over. And at the end of the day, NASA officials are highly sensitive to public perception, to the winds blowing one way or another within the legislative and executive branches, and countless other things. They know that they live or die by taxpayer dollars and the goodwill of their elected representatives, and they are not out to fool or mislead anyone. But this is the essence of how things must get done—not just with JPL or NASA, but with any government project of such a large scope. And if you think that this program, which ultimately cost $2.5 billion, is bad, check out military procurement sometime. Depending on how you parse the numbers, the entire MSL program's cost is about the same as the price tag of one or two B-2 stealth bombers, or the bailout of a really small savings and loan a few years back. In that light, it seems pretty mild.

There is another factor here, and that is the tyranny of the launch schedule. Mars is in the proper alignment for a launch attempt once every two years. So if your launch date slips, as MSL's did, you rack up costs as your completed spacecraft components sit and wait, slowly degrading from optimal condition, and people must be paid to continue taking care of business. MSL's launch window slid from 2009 to 2011 for a variety of reasons, and with troubling financial results.

As I said, this is a long and detailed story, and I aim to merely capture the essence of it here. What I do want to impart is an idea of the struggles that NASA and JPL face to acquire and retain funding for major projects, and to stay on schedule while at the same time restraining costs. Despite what you might hear on some AM radio rants, they are, in general, a frugal organization.

The best way to complete this part of our story is probably to look at NASA's own painfully honest assessment of the MSL program. NASA's Office of the Inspector General prepared a report dated June 2011, and, possibly to the surprise of some NASA detractors, it is a candid and sometimes-critical look at why things were slipping in terms of both cost and scheduling. I will include relevant portions and edit out others; if you want to see the entire document, steer your web browser to http://oig.nasa.gov/audits/reports/FY11/IG-11-019.pdf. It's interesting, if agonizingly detailed, reading.

NASA'S MANAGEMENT OF THE MARS SCIENCE LABORATORY PROJECT OFFICE OF INSPECTOR GENERAL REPORT NO. IG-11-019 (ASSIGNMENT NO. A-10-007-00) OVERVIEW

The Issue

The Mars Science Laboratory (MSL), part of the Science Mission Directorate's Mars Exploration Program (Mars Program), is the most technologically challenging interplanetary rover ever designed. This NASA flagship mission, whose life-cycle costs are currently estimated at approximately $2.5 billion, will employ an array of new technologies to adjust its flight while descending through the Martian atmosphere, including a sky crane touchdown system that will lower the rover on a tether to the Martian surface. Contributing to the complexity of the mission are the Project's innovative entry, descent, and landing system; the size and mass of the rover (four times as heavy as the previous Martian rovers Spirit and Opportunity); the number and interdependence of its 10 science instruments; and a new type of power generating system.

The primary objective of the Mars Program is to determine whether Mars has, or ever had, an environment capable of supporting life. In pursuit of this objective, the MSL rover—known as Curiosity—will assess the biological potential for life at the landing site, characterize the geology of the landing region, investigate planetary processes that influence habitability, and analyze surface radiation. NASA's Jet Propulsion Laboratory (JPL) is responsible for development and management of the MSL Project.

Due to planetary alignment, the optimal launch window for a mission to Mars occurs every 26 months. MSL was scheduled to launch in a window between September and October 2009. However, in February 2009, because of the late delivery of several critical components and instruments, NASA delayed the launch to a date between October and December 2011.

This delay and the additional resources required to resolve the underlying technical issues increased the Project's development costs by 86 percent, from $969 million to the current $1.8 billion, and its life-cycle costs by 56 percent, from $1.6 billion to the current $2.5 billion. If the Project is delayed to a late 2013 launch window, NASA's costs would further increase, at least by the $570 million that would be required to redesign the mission to account for differences in planetary alignment and the Martian dust storm season. [AUTHOR'S NOTE: All boldface emphases are mine, not in the original report. Also note that the actual launch delay was apparently announced in November 2008, not February 2009. It is not clear why the report gave the later date.]

In light of the importance of the MSL Project to

NASA's Mars Program, the Office of Inspector General conducted an audit to examine whether the Agency has effectively managed the Project to accomplish mission objectives while meeting revised cost and schedule projections. See Appendix A for details of the audit's scope and methodology.

Results

We found that the MSL Project has overcome the key technical issues that were the primary causes of the 2-year launch delay. Additionally, as of March 2011 all critical components and instruments have been installed on the rover. Project managers expected to complete integration of equipment by May 2011 and ship MSL to Kennedy for flight preparation by June 2011.

However, of the ten issues Project managers identified as contributing to the launch delay, as of March 2011 three remained unresolved: contamination of rock and soil samples collected by the Sample Acquisition/Sample Processing and Handling (SA/SPaH) subsystem and development of flight software and the fault protection systems. The resolution of these and other issues that may arise during final integration is likely to strain the already limited margin managers built into the Project's schedule to allow for unanticipated delays. **Moreover, since November 2009 this schedule margin has been decreasing at a rate greater than planned.**

In addition, approximately 1,200 reports of problems and failures observed by Project personnel remained open as of February 2011. If these reports are not resolved prior to launch, there is a possibility that an unknown risk could materialize and negatively affect mission success.

Finally, since the 2009 decision to delay launch, the Project has received three budget increases, most recently an infusion of $71 million in December 2010. **However, in our judgment because Project managers did not adequately consider historical cost trends when**

estimating the amount required to complete development, we believe the Project may require additional
funds to meet the 2011 scheduled launch date.

[. . .]

As early as May 2009, MSL's Standing Review Board expressed
concern about delays in development of flight software
and fault protection systems and we are concerned that
their development remains incomplete. As of March 2011,
the majority of the software needed for launch, cruise,
entry, descent, and landing was developed. However, the
software was not expected to be delivered until May
2011 and Project managers stated that work on software
required to operate the rover on Mars would be completed
after launch. In addition, as of March 2011, 13 of the 16
necessary fault protection related tasks had been completed and the remaining 3 were in progress.

**Because of technical issues related to these three
and other items, Project managers must complete nearly
three times the number of critical tasks than originally planned in the few months remaining until launch.**
As shown in Table 1, Project managers had planned
to have all critical tasks (except for Kennedy Space
Center operations) completed by April 2011. However,
when they revised the schedule in November 2010, that
date slipped by 3 months to July 2011. Furthermore, the
February 2011 revision shows that seven critical tasks
have been further delayed. Coupled with the decreasing
schedule margin described below, **we are concerned that
management may be pressured to reduce mission capabilities in order to avoid another 2-year delay and the
at least $570 million in associated costs.**

Accelerated Schedule Margin Decrease. To allow for
unanticipated delays, NASA routinely builds a margin
of extra time into project development schedules. We
found that for MSL this schedule margin has eroded at a
rate slightly greater than planned and that as of February 2011 only 60 margin days remained (see Figure 4).

When the launch was rescheduled in 2009, Project managers programmed 185 margin days into the development schedule. However, since November 2009 the Project has been consuming margin days more quickly than managers expected as a result of the number and complexity of technical issues needing to be resolved. **Although managers expressed confidence that the remaining schedule margin would be adequate to address the risks having potential schedule impact that they have identified, the rate of schedule margin decrease concerns us because the inherent complexity of the MSL Project increases the likelihood that additional issues will arise in final testing and integration.**

[. . .]

[Now comes the part where the report criticizes specific areas, of which I have included only a few:]
Project Management Did Not Effectively Assess or Prioritize the Risks Identified by the P/FR Process. Problem/Failure Reports (P/FRs) are generated when individuals working on a project observe a departure from design, performance, testing, or other requirements that affects equipment function or could compromise mission objectives. P/FRs may range from minor issues with negligible effects to potential "red flag" issues with significant or major effects, up to and including a loss of mission.

[. . .]

Project Funding May Be Inadequate. The Project achieved several important technological successes over the past 2 years, including delivery and acceptance of the actuators (motors that allow the rover and instruments to move), avionics, radar system, and most of the rover's instruments. However, Project managers did not accurately assess the risks associated with developing and integrating the MSL instruments and did not accurately estimate the resources required to address these risks. Consequently, the cost of completing development and the Project's life-cycle costs have increased.

In August 2006, NASA estimated the life-cycle cost for MSL as $1.6 billion. After launch was rescheduled for 2011, Project managers developed a new schedule and cost baseline for the Project, adding $400 million to complete development. Estimated life–cycle costs for the Project increased to $2.3 billion in fiscal year (FY) 2010 and to $2.4 billion in FY 2011. In November 2010, the Project requested an additional $71 million, which brought the total life-cycle cost estimate to the current estimate of approximately $2.5 billion. The extra money was obtained by reprogramming funds in the FY 2010 Mars Program budget, identifying additional funds from the Planetary Science Division in FY 2011, and addressing the balance in the FY 2012 budget request.

[. . .]

Conclusion. Historically, NASA has found the probability that schedule-impacting problems will arise is commensurate with the complexity of the project. MSL is one of NASA's most technologically complex projects to date. Accordingly, we are concerned that unanticipated problems arising during final integration and testing of MSL, as well as technical complications resulting from outstanding P/FRs, could cause cost and schedule impacts that will consume the current funding and threaten efforts to complete development and launch on the current schedule. Similarly, we are concerned that the limited remaining schedule margin may increase pressure on NASA to accept reduced capabilities in order to meet the approaching launch window and avoid another 2-year delay that would require significant redesign at a cost of at least $570 million or cancel the mission.

So there you have it . . . mostly. This is a very abbreviated version of just the report's overview; it goes on for another forty-two pages in excruciating detail. But that is how proper audits are done. Don't you wish that all government-procurement programs were this transparent? One of NASA's great challenges since the early days is that, being a civilian agency, everything it does is in the full glare of public scrutiny, the media limelight, and the squinty eyes of hostile congresspeople and representatives. The money that must be spent just to maintain transparency, to report on the proceedings, is in itself blinding. And while it contributes to accountability, dollars spent on audits are not spent on flying to Mars or to the space station. It can be highly frustrating, and it is never easy to be this self-critical, especially when so many others are ready to pounce on every word. "See? I told you this government bureaucracy if full of pork and waste!" and so forth, ad nauseam. Yes, there can be issues, and there may even be occasional waste (though you are very unlikely to see it when you visit JPL—trust me on this). But overall it's a pretty tightly run ship.

Perhaps most important: if the agency was not at the whim of two- and four-year election cycles, shifting administrative priorities, and home-state-driven representatives, *and* if it could actually plan a decade ahead (like China's space program does, to its everlasting benefit), American space exploration would be a different story. But that is not how our system works, and until it does, agencies like NASA that run programs that span multiple presidential administrations will likely have to estimate low and then suffer the consequences. Just try not to be too alarmed when the budgets begin to pass the estimates—because those estimates were probably not realistic to begin with.

The scope of the program had expanded and the subsequent delays had cost money. MSL had slipped by a full two-year launch cycle, and besides incurring more expense, it threw the life schedules of 480 participating scientists and countless engineers, programmers, technicians, lunch servers, and a large janitorial staff into disarray. The launch delay had a bright side too, however, as many of the critical issues that were dogging the engineers in particular now had another year plus to get ironed out. In the end, it may well have contributed to the mission's success.

CHAPTER 14

MY MARS

And now for some self-indulgence. I've been obsessed with space since early childhood, and with Mars just about as long. This began during the second act of the space race, the early 1960s. It was still not proven whether or not there were vast forests of lichen, leafy plants, or perhaps even evolved life on Mars. Admittedly, much had changed since the days of Percival Lowell. Others had made extensive telescopic observations, and few saw the canals or oases he and Schiaparelli did. Some used spectroscopes and found little moisture in the atmosphere, and still others determined that the air pressure there was about 1/1000 of Earth's. It was not a promising set of observations for those of us still hoping to find a Lowellian Mars.

The Mars of Lowell, Edgar Rice Burroughs, and Ray Bradbury was rapidly vanishing. My *Star Trek*–inspired visions of green-skinned Orion dancing girls was, reluctantly, evaporating with them.

Then Mariner 4 flew past the red planet and the Martian empire, whether plant or intelligent beings, was toast. What was left was a dry, forbidding expanse. I was not a happy camper.

Of course, this was also the era of Apollo, so there were other obsessions to be indulged. But Mars was never far from my mind. It was pretty certain by then that the Victorian vision of a Venus populated with riotous jungles and dinosaurs stomping across that steamy world did not exist, and ultimately the planet turned out to be a hellhole. And long before, the other planets had been discounted as places where beings could live—Mercury was too hot, and the gas giants—Jupiter, Saturn, Neptune and Uranus—were wholly inhospitable. And poor Pluto later lost planetary status, being demoted to a "dwarf planet." The solar system beyond Earth was rapidly becoming a lousy place for people—human or otherwise. Mars had joined the unfortunate, forbidding list.

When I was about ten, the year after that evil Mariner 4 forever ruined Barsoom, my father took me to Griffith Observatory in the hills above Hollywood. We spent the afternoon looking through the museum and seeing the planetarium show. Then evening came, and from the building's roof, Los Angeles spread out gleaming like a bejeweled carpet. It was a magical scene, but nothing compared with was yet to come.

Soon after dusk, the door to Griffith Observatory's telescope building opened. There were three green-patinated copper domes on the roof of the museum—one for the planetarium, one for a solar telescope, and the third for the astronomical telescope. The public was invited in—as they were every clear night except Monday—to view whatever was on the menu for the evening. The instrument was a twelve-inch-diameter refracting telescope that had been ordered from Carl Zeiss in prewar Jena, Germany, in 1930. It was among the finer instruments of its size for the time. The telescope's mount tracked the celestial motions with a rather-elaborate clock drive, electrically powered and with huge knife switches and bulky glass insulators, that strongly resembled something Dr. Frankenstein would covet. I liked it too.

Fig. 14.1. GRIFFITH OBSERVATORY: Built by the Works Progress Administration, Griffith Observatory opened in 1935 and included a planetarium, a science museum, and two telescopes—one solar and one celestial. Griffith Observatory has inspired generations of youth toward space and astronomy. *Image from klotz.*

A fine older gentleman named Gordon Mitchell was the telescope demonstrator that evening. He was very good with kids and made me feel like the most important person in the universe for a night. I was led up the wooden steps to the eyepiece, and there before me, against a deep-black field, was Mars.

As noted previously, the best telescopes yield a hazy, wavering red orb. In a smaller instrument, above the brightly lit and smoggy LA landscape, it was definitely suboptimal viewing. But it was still Mars—my Mars—and that first viewing has become a cherished memory.

I asked endless questions until my father took the cue and ushered me out, as there were other children, some surely as starstruck as I was, waiting in line.

When I was eighteen, I would return to the observatory as an employee—a museum guide. Two years later I was in charge of the guide staff. It was the best part-time job I could imagine while in college, and I stayed there for almost seven years. Many others have stayed far longer; some whom I shared that time with still work there—it's that kind of place.

As employees we had full run of the building, and we were put in charge after 6:00 p.m. I had keys to virtually everything, and the Zeiss telescope was freed up after we closed for the evening. The demonstrator from my youth, Gordon Mitchell, still worked there and would often turn the telescope to the object of our choice after hours. There we would stay, exploring the cosmos until the wee hours. It was fantastic.

Meanwhile, I struggled through classes in UCLA's astronomy department. Then, in 1976, Viking 1 was scheduled to land on Mars on July 20th. I did not have the pull or connections to get into JPL at the time, but I did live only a few miles from Caltech, which manages JPL. I suspected that they would be hosting a live video link (quite exotic for the time), so I headed down at the requisite hour and, um, found my way into the auditorium. As I recall, it was not open to the public, but that was not about to stop this space-obsessed young man.

I stood near the back of the auditorium with a hundred others for whom there were not enough seats. There was a large, movie-sized video projection on the screen onstage. The room fell silent as the final minutes of the descent from orbit were relayed from JPL's mission control, a few miles to the north. We all knew that the transmissions were delayed by close to twenty minutes, and that success or failure had already occurred. But it *felt* live, and to a person the attendees were swept up in the moment.

Down, down Viking went. Nobody had any real idea what awaited it in the Chryse Planitia region—the orbital mapping was just not good enough then. The area had the virtue of holding some geologic interest as well as appearing to be relatively safe, but there was really no way to know. There was a lot of guesswork involved in selecting landing sites at the time, and a thousand kinds of surface features—invisible from orbit in 1976—could destroy the spacecraft instantly. It was the era of the "Big Dumb Lander," as JPL'ers would later refer to the Viking surface probes, and it was anybody's guess how it would end.

A few minutes later, a safe touchdown was confirmed. Where the Soviet probes had failed just a few years prior, the United States had succeeded brilliantly, and in the year of the American bicentennial, too. It was just fantastic, as Rob Manning would later say—frequently—about his Mars missions. I couldn't agree more.

Immediately after landing, the cameras were turned on and began sending down the first image from Mars. Contrary to the expectations of many, it would not be a magnificent color panorama of the landscape, but a black-and-white shot of a footpad on Martian soil. The engineers wanted to make certain that the spacecraft had landed level, and on solid ground.

Strip by strip, top to bottom, the picture came in a line at a time, moving left to right. It was completed over the course of a few minutes. Though it was a monochromic image, the primitive video projectors ended up displaying it as purple. But I didn't care—I was looking at the surface of Mars!

The next day, the first color landscape shot was delivered, and what a photo it was. Newspapers and TV reports carried it across the globe. It was amazing—deep saturated red soil and bluish sky (the image was later corrected to reflect a more accurate salmon-colored skyline, to which a few snarky reporters attending the press conference responded with the likes of "What's it gonna be tomorrow? Green?" Some people just don't appreciate the difficulties of planetary exploration).

Over the next decade, the pursuit of girlfriends, a long trip around the world, and a career in television would pull me away from Griffith Observatory, the space program, and the red planet. But not for long. And I never forgot the Mars I was able to explore at Gordon Mitchell's telescope at the tender age of ten, nor that first image from Viking.

We now return you to our regularly scheduled chapters. Thanks for listening.

CHAPTER 15

HURTLING TO MARS

On November 26, 2011, at just past 10:00 a.m., the Mars Science Laboratory spacecraft departed Cape Canaveral, Florida, from Launch Complex 41. It's a pad reserved for unmanned launches, smaller and less elaborate than the one you were used to seeing the shuttle launch from in the past. The rocket was an Atlas V, and while there is one larger booster in the US arsenal (the Delta IV Heavy), the Atlas makes a heck of a racket when it departs. I watched an Atlas V launch in 2009 from about six or seven miles away with my thirteen-year-old son, and the look of shock and awe on his face was gratifying. Not as booming and grand as a shuttle launch, but plenty loud.

There had been the usual last-minute checks and tweaks before the MSL launch, but in addition, there had been some eleventh-hour drama beyond the norm. An issue with the drill—a potential short-circuit-causing defect—had been discovered, and Rob Manning and his team had quickly designed and implemented some last-minute fixes. It was a close thing.

The first stage burned as planned and dropped off to break up over the Atlantic. The second stage, called a Centaur, pushed MSL the rest of the way toward escape velocity. Both the Atlas and the Centaur are part of the old guard of American rockets, though the Atlas has been substantially changed since the early days (in fact, the first stage is now powered by—heavens!—Russian rocket engines). Both are reliable, proven designs, and both did their jobs that day.

MSL shed the Centaur once it was on its way to Mars. But before it cut loose, the spacecraft was set spinning at two revolutions per minute along its center axis for its long trip. This continually changes the side of the spacecraft exposed to the sun, equalizing temperatures across the structure. It also provides stability for the craft, in the same way that imparting spin to a football when thrown causes it to fly straight.

The entire spacecraft now consisted of the Curiosity rover and its landing rockets, enclosed in a cone-shaped aeroshell, and a cruise stage. The cruise stage was a ring-shaped unit that had its own rocket motors and power system. Its rockets, intended for course corrections, were small ones, using a nasty monopropellant (a fuel that does not require a separate tank of oxidizer like liquid oxygen) called hydrazine. Besides being highly toxic, hydrazine is, as you might imagine, highly reactive. I have heard that the engineers who worked at Grumman Aerospace in the 1960s, building Apollo's lunar module, used to show the new guys just how reactive mono-propellants were with a graphic demonstration. They would take the newbie outside, preferably in the winter, squirt some hydrazine onto a frozen bush, and watch it explode. If true, this exercise would have made the point with little discussion.

The spacecraft made the 352-million-mile crossing to Mars in just over eight months without mishap, then the cruise stage separated from the lander's aeroshell ten minutes before atmospheric entry. The spacecraft then used small rockets to cancel out the spin imparted when it left Earth. It was now flying blunt end first and heading directly for Mars. And here is where it got really interesting.

I'm sure you have seen the "7 Minutes of Terror" video that JPL posted online before the MSL launch. It was a nicely crafted and exciting piece dramatically illus-trating the entry, descent, and landing (EDL) phase of MSL. If you are one of the five or ten people with access to the Internet who have not seen it, treat your-self. Besides being highly informative, it's also fun. It will also introduce you to Adam Steltzner, the engineer in charge of the EDL sequence who became known as "NASA's Elvis" due to the pompadour he sports. Turns out he is also a highly entertaining informer.

The main point here is that the MSL spacecraft went right from interplanetary cruise into the Martian atmosphere like a bullet. None of this "let's go into orbit, pick a nice place to set down, and then land" stuff from the Viking days. From Pathfinder onward, all the missions had simply aimed the spacecraft at where Mars would be in eight to nine months and scored a bull's-eye. The orbital missions have it a bit better: when they reach Mars, they use a technique called aerobraking, in which they go into a large, lopsided orbit, dipping into the thin atmosphere over a period of months to scrub off speed and reach a nice, stable orbit. Not so the landers—they just barrel in and land.

When MSL reached Mars, it was traveling at over 13,000 mph. That's fast. The upper Martian atmosphere is 1 / 1,000 the pressure of Earth's and not dense enough to slow an incoming spacecraft anything like Earth's far thicker atmosphere slowed Apollo capsules or the space shuttle. A heat shield and parachute are not enough to slow a heavy lander to a gentle landing on Mars. Something more complex is needed.

Fig. 15.1. NASA'S ELVIS: Adam Steltzner, who gained fame in JPL's viral video "7 Minutes of Terror," was in charge of the entry, descent, and landing (EDL) phase of MSL's mission. Credited (along with "Mohawk Guy" Bobak Ferdowsi) with helping to make NASA "cool" again, Steltzner came to his career choice later in life after spending time in his twenties in a rock 'n' roll band. *Image from NASA/JPL-Caltech.*

Here's where I will start to go Tom Clancy on you—so prepare yourselves.

Anyone who has followed the MSL mission knows that it has a high degree of autonomy in its surface operations. It was the same story in space, especially in the final phases of the journey. JPL had defined the landing zone, or "landing ellipse" in their terminology, at about twelve miles long by five miles wide. The long axis corresponds with the direction of flight. The width is determined by any deviation from the straight line as the spacecraft comes in for a landing. It was a far smaller zone than ever attempted, and it would require far greater accuracy.

Curiosity was headed for Gale Crater. At its widest point, Gale was still twenty-five miles smaller than the MER's landing ellipse's long axis. And with Mount Sharp dominating its center, the strip of real estate deemed safe for landing was small indeed. This would require unprecedented levels of accuracy and control. A carefully controlled, guided entry through the upper atmosphere, driven by a relatively powerful computer, would be needed.

And MSL did possess a much more powerful computer than its predecessors. Though sluggish by the off-the-shelf standards of the time, it was the most modern, affordable, radiation-hardened chip available, and it was perfect for the intended set of tasks. It was called a RAD750 PowerPC chip, a descendant of the Motorola CPUs that Apple used in its G3 Macintosh computers around the turn of the twenty-first century. It was still ten times as powerful as those used on MER, and with the highly efficient and dense software written by JPL, it would do the job handsomely. For those who enjoy the numbers, it has 10.4 million transistors (MER had 1.1 million), runs at a maximum of 200 MHz (MER ran at 20), and has access to two gigabytes of flash memory. There are two identical units on board the rover, one primary and one backup.

For landing, the computer had a set of parameters and values preloaded and was capable of making many of its own decisions as it hurtled into the Martian atmosphere. It knew where Mars was in three-dimensional space (due to exacting and recently updated calculations provided by the ground), and it knew where *it* was in three-dimensional space (via Earth-based measurements and its own onboard inertial measuring systems). It also had multiple thrusters mounted to the outside of the aeroshell to give it maneuvering capability to make any needed corrections as it headed in. As the spacecraft got closer to the surface, the onboard computers were able to keep the landing zone's location in the navigational crosshairs and adjust the trajectory in real time. MSL was now headed in the right direction; all it had to do was track the target and survive the landing.

The landing software included a provision for a gliding entry. The fifteen-foot heat shield was based on the old Viking design. It was not a pure cone, but had an offset bulge on the bottom, making it, in effect, a low-efficiency glider. As it sped toward the surface, it was able to "surf" on the thin blanket of air, scrubbing off some of its blinding speed and allowing it to steer toward the landing area. The maneuvering thrusters helped to keep it aimed true. There were also eight tungsten weights, heavy gray cylinders, mounted inside the edge of the structure.

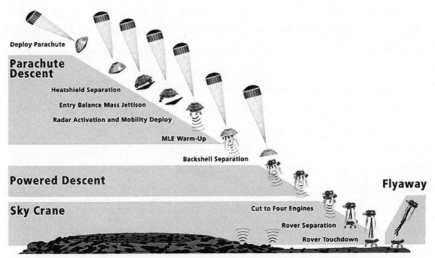

Fig. 15.2. SEVEN MINUTES OF TERROR: This chart shows each phase of the entry, descent, and landing (EDL) sequence. Top left is atmospheric entry, bottom right is landing. *Image from NASA/JPL-Caltech.*

As the heat shield began to glow white hot, it burned away, a process called ablating. This is how most heat shields work—heat is carried away by the burning materials as they are shed. The trick is to make sure that it burns evenly, so that it will maintain its aerodynamic shape.

The math needed to glide properly was based on work done well over fifty years earlier for the Apollo program. The capsules returning from the moon also needed to glide, or, more precisely in their case, skip along the atmosphere as they came home to reduce speed. While Martian air is much thinner than Earth's, the same principles applied.

To glide effectively, the spacecraft had to shift its center of gravity. This was accomplished by ejecting the first two weights, each 170 pounds. The craft pivoted to aim the offset heat shield in the proper orientation for gliding and steering. Fine adjustment was accomplished by the small rocket thrusters.

By the time this portion of the atmospheric entry was complete, MSL had slowed to Mach 1.7, or about 1,300 mph. It had lost a lot of speed from the roughly 13,000 mph it was traveling when it reached Mars, but it was still going way too fast. Six more tungsten weights, each weighing fifty-five pounds, were then tossed away, bringing the spacecraft back to an even keel. At just over six miles from the surface, the parachute deployed. The parachute had given the engineers fits, as it had ripped repeatedly during testing. And working with cloth is a nightmare in these kinds of tests. Metal is nice and predictable—bend or pull it this much, and it will weaken, this much more, and it will break. Cloth becomes a somewhat-unpredictable mass of small fibers that is hard to quantify, and some people were still nervous about the parachute's performance. For Martian landings, only Viking's parachute had been of a similar size, but that spacecraft was only about half as heavy as MSL. There were just so damn many variables. But they had tested the fifty-foot chute in the nation's largest wind tunnel, and it had held at speeds in excess of this, so it *should* work. . . .

It did. The parachute deployed, then widened as it un-reefed (a gradual releasing to full width), slowing the lander. As it settled into a mostly vertical fall, the heat shield, its work done, was dropped. The MARDI camera started to record the final phase of the descent, and would provide dramatic imagery of the ever-nearing landscape. It was then that the Mars Reconnaissance Orbiter tried for a picture of MSL during its final few minutes in the air, aiming at what the MRO controllers had predicted would be the right place. Amazingly, it worked, and provided a stunning photo of a beautiful, billowing parachute.

Two minutes to go. Flight controllers back on Earth were still monitoring signals that indicated that the spacecraft was outside the atmosphere due to the long delay for the radio signals in reaching Earth. MSL was way ahead of them, flying on its own.

With just over a mile to traverse and still traveling nearly 200 mph, the rover and its rocket pack dropped free from the remainder of the aeroshell, leaving it and the parachute behind. This was the terminal phase of the landing, and the scariest for its creators.

It was time for sky crane to go to work.

CHAPTER 16

SKY CRANE

"**W**hat the f—" were my timeless first words upon encountering sky crane. That was my inner voice. The outer voice muttered something like, "By golly, that's interesting." I would later learn that many in the planetary-science community, and beyond it, uttered something more akin to the former than the latter. The approach was *weird*.

As a person who had written about, covered, and tracked the US space program for a couple of decades, this struck me as a decidedly different approach to landing on another world. Viking had landed like a proper, macho lander—upright, rockets blazing after a parachute-slowed entry into the Martian atmosphere. With Pathfinder, things began to get weird—the second US mission to land on Mars did so covered in fabric beach balls, bouncing dozens of feet into the thin Martian air before rolling to an eventual stop. The Mars Exploration Rovers, Spirit and Opportunity, arrived in a similar, somewhat–Bugs Bunny style, bouncing over a dozen times, then rolling to a stop as their airbags deflated.

And then there was the Mars Science Laboratory. How the heck do you land something that large, and heavy, and that is traveling at interplanetary speeds?

"It was thinking out of the box. In fact, we threw away the box. We were literally going through all possible ways to land this machine, trying to imagine every possible configuration, whether it made sense or not. You do this until you bump into one that's so crazy that it makes sense." That's Rob Manning once more. Again, he was at the nexus of figuring out how to set down a machine on Mars while staying within budget. Again he led a small team of equally brilliant people to work through the problem. And again, they came up with something that made people sit up, swallow hard, and strain their minds to get their head wrapped around it.

I realize that I left you hanging a few dozen feet above Mars in the last chapter. But before we look deeper into how sky crane works, I want to share its origins.

Nobody can do that in a more entertaining way than Manning, so let's hear the rest of the tale from him.

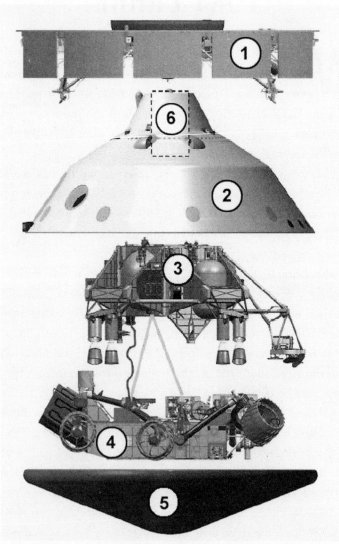

Fig. 16.1. MSL'S PIECES: This graphic shows each component of MSL: (1) the cruise stage (discarded before atmospheric entry), (2) the aeroshell that protected the rover and descent stage during entry, (3) the decent stage/rocket pack, (4) the rover, Curiosity, (5) the heat shield, and (6) the parachute. *Image from NASA/ JPL-Caltech.*

"So NASA had just experienced two embarrassing failures with the Mars Polar Lander and the Mars Climate Orbiter in 1999. Before these missions failed, I was leading the effort to build an oversized version of the Mars Polar Lander to carry a rocket and a sample-return rover on the top deck. It was supposed to land in 2003 and 2005, there were going to be two of them. . . ."

As you will recall, the Mars Polar Lander and Mars Climate Orbiter missions were both part of the "faster, better, cheaper" era. And, unlike Pathfinder, both failed. It was clear to NASA and JPL that the next few missions had better work. The MER rovers landed just fine, but the landing system for the big MSL mission was being discussed before MER's successes, and the engineers were understandably a bit gun-shy.

Back now to Rob and his big, oversized post–Mars Polar Lander mission: "So these [new] landers were supposed to collect samples, rocket them off towards Earth, and by 2011 we would have samples from Mars. But we previously hadn't spent a lot of time trying to figure out if something that size and that scale would work, so my team at Lockheed was struggling with it. We were trying to make a payload [rover] that could drive off the top deck of the lander." Remember that Pathfinder and MER had used landers that folded open, the sides becoming ramps for the rovers. That worked well for smaller machines, but you start getting up into the big leagues—say, a bulky two thousand pounds or so—and this approach does not scale up well.

Rob continues: "The ramps were getting huge, the rover was getting huge, and the lander was so big by now we realized we did not have rock clearance." You need to be able to land on rocky surfaces when sending machines to the Martian surface, and at this scale they were having trouble guaranteeing clearance of larger boulders. "Trying to find a place on Mars that is rock free is still a big challenge." Indeed, it's like trying to find a part of the Atlantic that is dry. Using the vastly improved images from Mike Malin's orbiting cameras, however, they were indeed able to find areas that looked *fairly* safe. But that presented a new problem. . . . "The geologists want you to go to rocky places so that they can investigate the rocks, so to accommodate them we were struggling to get that big lander down safely. Then these two failures happened, and we said to ourselves 'Okay, we have some serious issues here getting things to Mars safely, irrespective of why Mars Polar Lander disappeared. Let's go back and rethink landing on Mars.' So we did."

Two failures in a row is enough to make even the most enthusiastic, optimistic engineer a bit cautious. It was time to start with a clean sheet of paper. The traditional approach to landing a big, heavy rover was, at the time, very much like the Viking lander—use a large descent stage that lands on legs with rockets firing. The problem was not so much the lander, though, it was the rover—putting a one-ton rover on top of a lander creates problems. The dynamics get very wonky, and it's a bit like trying to balance a bowling ball on a broomstick—the center of gravity is very high and the system becomes unstable. So they looked at all the ways they could think of to get around the balance problem.

"A small group of us got together, and we fiddled with all these permutations," says Manning. "One of them was a variation on Mars Pathfinder. We would have to replace the solid rocket motors with liquid rockets that could be throttled. Then we would lower the lander on ropes, and we wouldn't need big airbags anymore— if you control it well enough, you landed using the ropes with no airbags at all. You just land your rover right on its wheels." This did away with the issues of a rover sitting on top of a landing stage, but was, to say the least, an unorthodox approach. Manning remembers his first feedback on the idea: "The three of us were pretty excited about this. We started talking to some of the dynamics-control guys [the engineers involved in flying the lander to the surface of Mars] and they said, 'Look—we just don't know enough about the dynamics of landing a system this way . . . we can study it later, but let's not do it now—it's just too crazy.' So we put that idea off to one side." What the engineers said when Manning and crew left the room is not recorded. . . .

Sky crane was sidelined, but not for long.

Manning continued: "MER had not happened at that point. The whole idea of landing a rover on its wheels with the control system overhead, suspended by ropes seemed a little out there. So we had another idea which looked more promising called the pallet lander. It's a version of the lander that's more crushable."

When any machine lands on another planet, the challenge is to lose energy, in this case downward speed, before or when you touch the ground. If you have lots of money and big, powerful rockets, you can do it the way Viking did—slow yourself on those big, expensive rockets until you make ground contact, on shock-absorbing legs. Or you can use other ways to slow down, to absorb the shock. The pallet system would allow the whole landing stage to crush itself in a controlled fashion, which would result in the absorption of a lot of energy (and, in effect, speed reduction) and get the top deck, and the rover, closer to the ground.

"So the platform lands and crushes itself. It is very stable, and you can actually drive off this flattened lander, which is more conforming to the surface." Again, thinking outside the box—in effect, destroying the spacecraft's lower stage as it lands to give the rover a knee-level roll-off to the ground. Brilliant. "We studied that for a couple of years. At that point the only lander mission we had put together besides Viking was Pathfinder. Mars Polar Lander was dead and the Mars Exploration Rover mission hadn't started."

Manning was moved over to the MER program to make his airbags and rovers work. But MSL marched on, says Manning, and a new factor entered the discussion—those bigger, variable engines we talked about that landed Viking so successfully. "A very small team was still looking at ways to make the crushable-pallet system work. That's when we got money to develop the throttleable engines."

Manning may have moved to MER, but the heavier, more capable MSL rover was always in the back of his mind—lurking, waiting, taunting. He continued: "We spent three years getting MER off the ground and to Mars, but while we were developing MER, I realized that we were learning a lot about the dynamics of using rope up on Mars. That idea that we had—the precursor to sky crane—is not as crazy as people thought." It was time to campaign a bit. "I tried to talk to the current manager of MSL—which was then called the Mars Smart Lander—and he was interested. They got some workshops together and started having discussions about this option."

The workshops accomplished what they were intended to do—collect a bunch of smart people, look over the problem from all angles, then assault each other's thinking until you're pretty sure it can be done the way you're discussing. Remember the "hazing" John Grotzinger discussed regarding sedimentation on Mars? This was similar. "So after a few workshops between 2000 and 2003, we decided to should look into this approach more carefully." The MSL team looked at the dynamics of a rover slung *below* the lander. "Since many of us were still busy on MER, we would come swooping in and then talk about the MSL landing system, then go back over to MER." It was like drive-by brainstorming—spend a few hours on the Opportunity and Spirit rovers, sneak off to meetings or working sessions about MSL, then get back to the MER project. But soon, MER landed and there was more time to consider options for MSL. They decided that the nascent sky-crane system was their best option. "Now the hard part was convincing the rest of the world," Manning says with a wry grin.

"There were a lot of levels to convince. The next was JPL upper management, then the outside world. So we brought in a lot of independent people." These included experts from a McDonnell Douglas rocket project from the 1990s called the DCX, Sikorsky Helicopters, people who worked on the Viking and Apollo programs, and many others. "We even had Harrison Schmitt, an astronaut from Apollo 17," Manning remembers. "So we were going through the design, and everyone is poking at it, poking at it, and poking at it. We then took their recommendations and went back and worked on it for about six months."

They had what could be politely called "nagging doubt" to overcome. Manning calls it the "laugh test." "Their first response early on was: 'Are these guys serious? They seem serious!' The biggest question was, why would you go to something so outrageous if you have all these other ways you've done it successfully before?" The reviewers all knew about Viking, Pathfinder, and MER, so they understood landing with rockets on legs and, having overcome nagging doubt once before, the bouncing airbag method. But now . . . this? "So we had to explain for quite some time what the weaknesses were for all these other approaches. Sky crane was the only one that did not seem to have an Achilles heel. It seemed to work . . . at least on paper."

The design was indeed fascinating on paper, but it left a lot of problems to be worked out. First they needed a way to slow the spacecraft to a crawl before sky crane could even begin to do its job. That required parachutes and rockets. Once they had slowed their descent to a (fast) walking pace, "We needed to communicate between the rover and the descent stage, and we needed a way to lower the rover down."

That's where the ropes he spoke of earlier came in. And the ropes are what engendered most of the horrified looks—or giggles—from onlookers.

"At this point we were a bunch of true believers," he recalls with conviction. "We needed not just management but outsiders who really knew the area to believe in it as well. The hard part now was going all the way to the top," meaning NASA *senior* management. Here they got lucky. The NASA administrator at the time was Michael Griffin. "Mike was a quintessential aerospace engineer, probably the most overqualified NASA administrator and history in that light. He's got more sheepskin than any other. So once we got to that level, the only person we really had to convince was him—Mike Griffin." Manning is clearly fond of the man. "He's not a delegator, he's a 'show me' kind of person. The way he worked was, 'If you can convince me, this is going to happen. If you can't convince me, then it's not going to happen.'" Adam Steltzner, who had taken over management of the entry, descent, and landing part of MSL, gave the presentation at NASA headquarters in Washington, DC. Griffin listened coolly at first, then with increasing interest. Manning continues: "Griffin finally said, 'Well if anybody can do this, you can.' I'm pretty sure he was thinking this is not how *he* would land on Mars; he might even have been thinking it was a bit Buck Rogers. But he let us do it. I'm sure Cadwell Johnson [of the original Pathfinder review group] would've said, 'Hey, it's supposed to have four legs you jerks!'"

So they had NASA's top vote in hand. Of course, at the same time that this top vote cleared the way to move ahead, it also meant a lot of pressure—the administrator, NASA's de facto CEO, is watching and has a personal investment in the project. And, he's a senior aerospace engineer—there's nowhere to hide.

They had permission. They had a budget—for now. Now they had to make it work.

I'm not an engineer. I don't have an advanced degree in the hard sciences. And when I saw sky crane, just a few years after JPL's two Mars missions had augered-in back in 1999, I was stunned. There appeared to be an awful lot of potential points of failure in this system—many dozens of explosive bolts (pyrotechnic fasteners that release when they explode), any one of which could decide not to work. There were the throttleable rockets on the backpack—wasn't Viking the last time they had built those? It also had to navigate to a very small zone in the middle of a deep crater—and do so mostly autonomously. And then there were those damn ropes.

Let's talk now with another engineer, Al Chen. We met him in chapter 2, the night Curiosity landed. His soft voice, which only really showed some emotional chroma when he announced touchdown, was the one you heard that night. He's an easygoing fellow in his late thirties who lives in the next community over from JPL in La Crescenta, is married to a woman who also works at JPL, and has three kids, aged seven, four, and just under a year, when we spoke. That's a full house—as a father of one, I can't even imagine. At any rate, Chen seems the type who would handle three kids with grace, kindness, and nary a harsh comment in sight. He picks up the sky crane saga, discussing why the cords between the descent stage (rocket pack) and the rover were a bit of a nightmare. In short: "Having things connected with soft things that are sometimes elasticky is scary. We don't need to have [a] hard structure connecting things to understand how it works, but it's nice." It's a bit like it was with the parachutes: when you move from dealing with metal parts and plastic circuit boards to cloth parachutes and springy ropes, the engineering numbers get soft—very soft. Testing becomes more critical than ever because the error margins are wide—prediction is almost as much about instinct as engineering. It did help that they had a fair amount of data returned from Mars by previous landings, but none of it modeled anything like this.

So what did he think when he first saw the design for sky crane? "This can't work. There's so many moving parts . . ." he recalled with a grin. "You definitely begin to feel like, when you take a step back several times in the process, 'What have we done to ourselves here?' We bit off so much new stuff on Curiosity that I'm surprised we talked people into letting us do it." I asked for examples. "Well, let's see: first guided entry on Mars, biggest planetary vehicle, biggest lander, and, of course, what is this new crazy landing system? There's a lot going on there, but when you look at it, does it hold together?"

Then, after the initial shock wore off, he reconsidered. "It was a natural evolution—a lot of the folks [who] worked MER and MSL were Pathfinder folks at the beginning," Chen observed. "Pathfinder in many ways is the opposite of what MSL is. On Pathfinder, they tried to do almost everything they could mechanically—very little guidance-navigation control. If, on the other hand, you turn around and look at MSL, it had almost gone the other way completely. We're trying to control everything."

Well, any father of three little kids knows how that works out in most cases. It's nearly impossible . . . but with Curiosity they came damn close. The guided-entry part of the mission—a new capability—was of course critical, but also doable. Mike Malin's orbital imagery was sufficiently high-resolution that it could identify a lot of potentially rover-murdering situations on the ground. Chen recalled, "We could see almost every rock that was a problem. With guided entry we knew we could track our lift and put the lander in a tight spot, literally on a very tight spot—considering that the walls of the crater were as high as mountains." In short: "Between sky crane and guided entry, we had to make very few compromises."

So using guided entry, loading the computer with landing parameters including hazards, and identifying the relatively small area within Gale Crater in which it was supposed to land provided the engineers with a modicum of comfort. But there was still that final seventy feet to deal with.

"An interesting part of sky crane for me is this," he continued: "Recognizing that

DAYBREAK AT GALE CRATER: This NASA-created image shows the terminator just crossing Gale Crater *(center top)*, with the mountain in the middle. With vast geological diversity, Gale Crater provided a worthy target for Curiosity. However, reaching it—with a narrow landing zone between the crater wall and Mount Sharp—would be a huge challenge. *Image from NASA/JPL-Caltech.*

TUCKED IN: Curiosity as seen at JPL. The rover and descent stage are folded together and tucked into the aeroshell. The bottom will be covered by the heat shield, and the assembly will then be mounted atop an Atlas V rocket for launch. *Image from NASA/JPL-Caltech.*

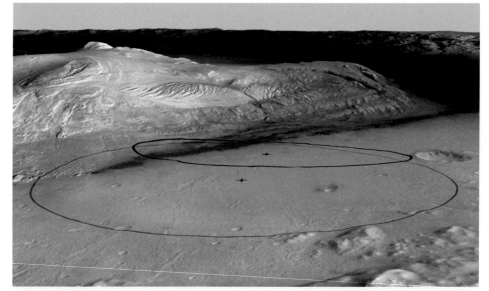

LANDING, REVISED: As the mission specifications changed, so did the landing ellipse, or acceptable range of error. The fainter line represents the old landing ellipse; the dark oval, the newer, far smaller one. Constant improvements in software and engineering meant a more accurate landing. *Image from NASA/JPL-Caltech.*

GALE CRATER: A wide overview of Gale Crater. This is a composite image created from data supplied by the orbiters. Clearly visible are the sharply defined edge of the crater rim and the smoother central mountain, Mount Sharp, rising about eighteen thousand feet into the Martian sky. *Image from NASA/JPL-Caltech.*

THE FOOTHILLS: Curiosity's ultimate target: the foothills of Mount Sharp. The sedimentary layering can be clearly seen. The rover will have to negotiate the winding pathways between the hills; it will be treacherous driving and slow going. With luck, Curiosity will last long enough to move beyond the mountain when its work there is done, which may take a matter of years. *Image from NASA/JPL-Caltech.*

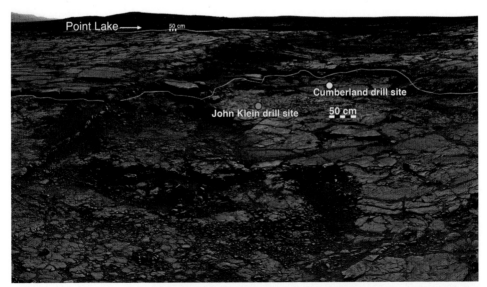

DRILL SITES: The two locations where Curiosity was sent to drill are seen in this image of Yellowknife Bay. John Klein is marked with the green dot, Cumberland with the yellow. The "50 cm" scale bar near Cumberland equates to about twenty inches. The foothills of Mount Sharp are seen in the distance to the upper left. *Image from NASA/JPL-Caltech/MSSS.*

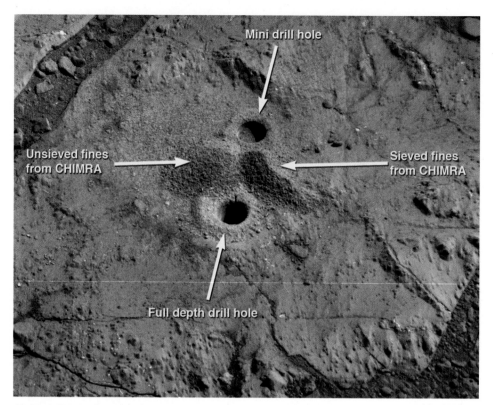

Mini drill hole

Unsieved fines
from CHIMRA

Sieved fines
from CHIMRA

Full depth drill hole

GRAY MARS: Finding that the soil on Mars is not an oxidized, rusty red just below the surface was a welcome development. It means that there is a better chance for life to have existed in the past on Mars. This image shows the John Klein test drill hole *(upper center)*, the main sample drill hole *(bottom center)*, and two small mounds of material dropped by CHIMRA after it separated the fine particles it needed for the instruments aboard Curiosity. *Image from NASA/JPL-Caltech/MSSS.*

OPEN WIDE: This is the sample funnel for CheMin with the door flap open. The mouth to the funnel, covered with a screen mesh, is about 1.5 inches across. *Image from NASA/ JPL-Caltech/MSSS.*

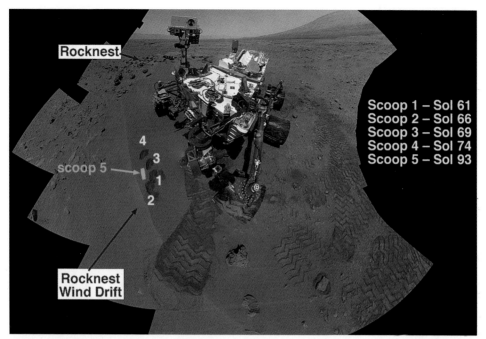

Scoop 1 – Sol 61
Scoop 2 – Sol 66
Scoop 3 – Sol 69
Scoop 4 – Sol 74
Scoop 5 – Sol 93

Rocknest

scoop 5

Rocknest
Wind Drift

HOW LONG? Robots work more slowly than people do. How long, then, does it take to sample one area? In this case, which was the first set of soil scoops taken by Curiosity and also allowing for a learning curve and system cleaning, it took just over a month. *Image from NASA/JPL-Caltech/ MSSS.*

OVER THE SHOULDER: Mount Sharp is seen behind the rover as the Mastcam looks over Curiosity's "shoulder." The nuclear power source, or RTG, is seen looking like a thorax with fins, angled up off the back end of the rover to the right. *Image from NASA/JPL-Caltech/MSSS.*

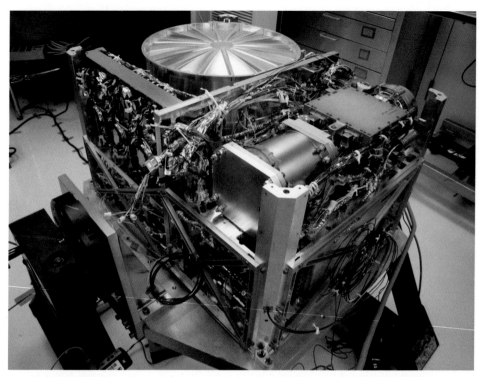

PORTABLE LAB: This is the SAM, or Sample Analysis on Mars instrument. About the size of a small oven, SAM comprises what would have been a medium-sized university lab just a few years ago. *Image from NASA/JPL-Caltech.*

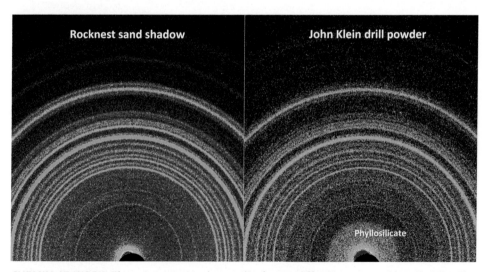

Rocknest sand shadow

John Klein drill powder

Phyllosilicate

CHEMIN AT WORK: These images are the result of x-ray diffraction tests within CheMin. The image to the left is generated from the windblown sand from Ripple at Rocknest, and the image to the right is generated by the drill sample from John Klein. The differences are subtle, but to the experienced eye, it is clear that the John Klein sample contains a lot more clay, which is water-formed. *Image from NASA/JPL-Caltech/Ames.*

HARD AT WORK: The drill is seen here working at John Klein. The bit is just above the small gray patch, and the locating/stabilizing posts are to the left and right of the bit. The DRT metal rock brush is visible to the far right, and MAHLI is the gold-colored circular object in the upper right corner. *Image from NASA/ JPL-Caltech/MSSS.*

FIRST TARGET: The rock Jake, named after deceased JPL engineer Jake Matijevic, was the first rock targeted for examination by the ChemCam (the red dots), as well as "contact science" with the MAHLI instrument and the APXS probe (the lightened circles). Jake turned out to be of a rock type previously not seen on Mars, known on Earth as a mugearite. *Image from NASA/JPL-Caltech/MSSS.*

TOUCHDOWN: This artist's rendering shows the moment of contact between Curiosity's wheels and Mars. Within seconds, the rover will send the "fly away" command to the descent stage hovering above and will release the cords, as well as the wired umbilical, allowing the rocket pack to fly off and impact elsewhere. *Image from NASA/JPL-Caltech.*

LIKE A JACKHAMMERED SIDEWALK: When Curiosity reached Hottah, it was a revelation. "Hottah looks like someone jackhammered up a slab of city sidewalk, but it's really a tilted block of an ancient streambed," said John Grotzinger. It was the first conclusive proof of flowing, hip-deep water on an arid planet. *Image from NASA/JPL-Caltech/MSSS.*

5 cm

TEST, TEST, TEST: Curiosity's parachute was a big challenge, as it kept ripping during development. Seen here is a test in Moffett Field's huge wind tunnel, the largest in the world, in Northern California. *Image from NASA/Ames.*

SPOTTED: Aiming purely by calculations, controllers of the Mars Reconnaissance Orbiter caught this image of MSL descending to the Martian surface. The rover has not yet been released from the aeroshell. *Image from NASA/JPL-Caltech/University of Arizona.*

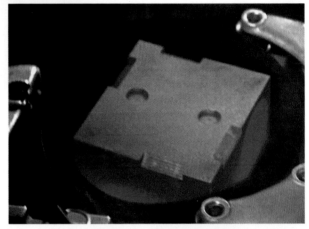

HOT STUFF: The plutonium fuel source is seen being placed inside Curiosity's RTG power source. It has a half-life of fourteen years and should last longer than that, if the Voyager spacecraft, still operational after thirty-seven years, are any indication. *Image from NASA/JPL-Caltech.*

JOY: Mission control erupts into jubilation as word comes back of Curiosity's successful landing. Rob Manning, MSL's chief engineer, is seen to the center left. At the top, turned away from camera, John Groztinger high-fives Pete Theisinger, MSL project manager. *Image from NASA/JPL-Caltech.*

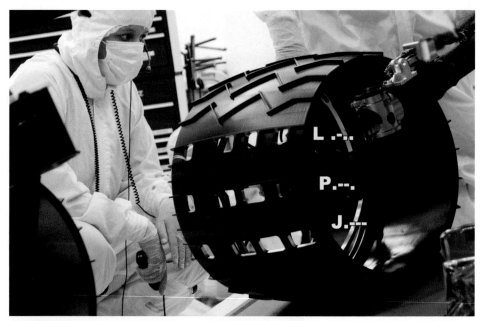

IT SAYS "JPL": A technician inspects one of Curiosity's six wheels. When the scientists wanted a way to count wheel revolutions, someone suggested embossing "JPL" onto the wheels. This idea was reportedly shot down by NASA, so instead the designers cut Morse code into the wheels . . . code for "J . . . P . . . L." *Image from NASA/JPL-Caltech.*

THREE FACES OF MARS: This image of Mount Sharp shows three ways in which images of Mars are viewed and interpreted. On the left, a raw, uncorrected version as it comes from the cameras. At the center, the image has been color corrected to what it would look like if you were standing on Mars. On the right, the image has been "white balanced" in order to see what the formations would look like if seen on Earth. This last technique is particularly useful when comparing Martian geology to formations found on Earth. *Image from NASA/JPL-Caltech/MSSS.*

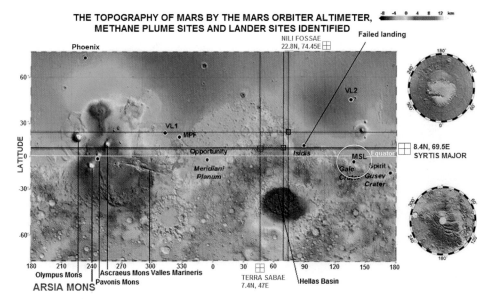

THE TOPOGRAPHY OF MARS BY THE MARS ORBITER ALTIMETER, METHANE PLUME SITES AND LANDER SITES IDENTIFIED

WHERE THEY WENT: This topographical map shows the landing sites of all US missions to land (or attempt to land) on Mars. It also shows topography: blue = lower, red = higher. Curiosity's landing area is circled to the right. *Image from NASA/JPL-Caltech.*

MARTIAN SELFIE: Curiosity shot this self-portrait over the course of many hours. The MAHLI camera on the arm had to be maneuvered into dozens of positions, then the images were stitched together to eliminate duplicate parts of the image. The final result is a flawless portrait of the rover after it had taken soil samples at Rocknest. The scoop trenches can be seen to the lower left. Mount Sharp is to the upper right. *Image from NASA/JPL-Caltech/MSSS.*

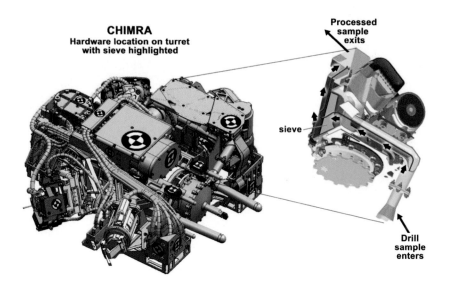

CHIMRA
Hardware location on turret
with sieve highlighted

Processed
sample
exits

sieve

Drill
sample
enters

HOW HARD CAN IT BE? To get a rock or soil sample on Mars, that is. Pretty hard, as it turns out. This schematic shows the CHIMRA, the part of the turret that separates and processes the sample before it is deposited into the instruments inside the rover. CHIMRA sits beside the drill, collecting rock powder from the bit or soil from the scoop and then running it through a system of sieves to get ever-finer grains. The arm must be moved into a whole series of positions to process the sample. *Image from NASA/JPL-Caltech.*

THE GRAIL: In the distance, Mount Sharp looms over the floor of Gale Crater, which Curiosity is slowly crossing. This image shows little of the dramatic terrain in the foothills (it's too far away), but the dark bands near the center indicate sand dunes that must be crossed en route—always a nerve-racking time for Mars rovers. Curiosity should begin working its way through the foothills by the end of 2014. *Image from NASA/JPL-Caltech/MSSS.*

once we get up to a mass like Curiosity has, we just can't do these energy-absorption systems we used before." No airbags. "So we should then flip the paradigm—instead of hitting hard at several meters per second, up to 10 meters per second, on the airbags, we should touchdown softly. If you touchdown softly enough, you don't have to worry much about breaking, it doesn't matter how big it is."

So bigger can be better after all. "Once you get to a rover this size, there are some advantages. If you touch it down softly enough, you can touch it down within the loads that are expected of it when it's driving around." It's already designed for driving over and falling off rocks if necessary. "So it's tough, resilient, and strong. If you touch down slowly enough, you already have a landing system—you have your own landing gear right there." The wheels and the suspension system would be its landing "legs." "It's designed to conform to the slopes, to deal with rocks; that's everything you also want in a landing system. Something that is very stable and will tackle whatever it has to deal with."

There is one more huge advantage of using the rover's own wheels as a landing system . . . "We don't have to drive it off a lander! A lot of people were still biting their fingernails when the MER landed because they were worried that we wouldn't be able to get the rover off the lander." The airbags, once deflated, always threatened to snag the rover as it departed. If the snag was too profound, the rover would have stayed right there, which would not have made it much of a rover. "With sky crane you don't have to worry about that as much."

"One technology that really made sky crane possible was good throttleable engines—like the Viking engines—that allowed us to get down and control how fast we were going with precision." This is why they resurrected the 1970, vintage rockets that landed Viking—they were a good design that, with a bit of improvement and updating, could land MSL right on its wheels. Past rovers—Pathfinder and MER—had used solid-fueled rockets to slow them. These are an evolution of the same skyrockets that you see on the Fourth of July. They burn a solid mass, with a fixed rate of thrust (and, of course, we hope that they do not explode). And once lit, they burn till they are done. If you know in advance exactly how much thrust you need, for exactly how long, they are fine. But with MSL's computer-aided navigation to the surface, responding to the ground-ranging radar, they needed something they could control. The amount of, and the duration of, thrust was critical. At the very least, these rockets needed to be able to throttle-up at the last moment, slowing the rover to almost a hover, when sky crane deployed and the rover was lowered from the rocket stage down to the surface. Every parameter was critical and would need to be handled with a high degree of precision.

"With those rocket engines we had everything we needed to get there. We had the landing gear, we had engines that we could control to slow to low velocities, and we can tell that we are getting to low velocities with the radar. Those were all the key building blocks of sky crane."

Chen smiled with satisfaction. "There was a rational reason for everything we did. Often we ended up flying systems that are a giant pile of compromises. We made very few compromises on sky crane. We got the engines we wanted that were very well controlled. We got the landing system that we wanted, the very capable high ground clearance rover with great landing gear. For the most part, we got the radar we wanted, it performed very well, told us our altitude and velocity right down to NASA precision."

So, to paraphrase, the JPL Mars crew had the Six Million Dollar Man (look it up if you are under forty-five): they had the technology, they made it better . . . and it was ready to assault the red planet in a bold and daring fashion. Now they had to do it.

I've already rattled off a narrative of the evening as seen from the media center at JPL, where I was fortunate enough to be stationed. It was an amazing evening from that vantage point. But Chen was at his console, running the show. Adam Steltzner was pacing like a predatory animal behind the row of consoles, keeping tabs on everything he could see, making decisions. Rob Manning was in the back row, watching his team's machines execute the exacting functions needed to land safely. John Grotzinger and a few other scientists had stationed themselves wherever there was spare room. JPL's director, Charles Elachi, sat in back with Charles Bolden, the NASA administrator.

Talk about *pressure*. The fact that the week leading up to the landing had been extremely quiet (in engineering terms, anyway) made Chen even more nervous.

"As the operations lead, I was responsible for essentially laying out the plans for the hundreds of people in the week prior to entry and landing. We do all these operational readiness tests and prepare for all sorts of crazy things. Then, the final week leading up to it was quiet. We were prepared for all contingencies. We were working around the clock, looking for problems, and we just didn't see any. It was kind of eerie. We were sitting around and looking at each other a lot, we ran a bunch of scenarios, and we still did all the work that we knew we should do, but in the end, we didn't have to do. Everything went like clockwork.

Fig. 16.2. SKY CRANE DEPLOYING: The main phases of sky crane involved the spacecraft being slowed by the descent stage (rocket pack), then the Curiosity rover being lowered by nylon cords. The descent stage was then cut free to fly away. Technical note: unlike this artist's rendering, the wheels were actually deployed downward by the time the rover was being lowered. *Image from NASA/JPL-Caltech.*

"We knew that cruise guidance had put us right on target. We didn't have to deal with *that* whole set of potential problems. I could throw out that part of the operations playbook. We were right down the middle. It's almost like something we didn't practice much—being right down the middle, right on the money. So this was kind of weird." He left the control center when off-shift, trying to distract himself. "Sitting in the media room, going next door, playing a game, trying to take a nap. In some ways, I wished there was something to *do* when I was there. At least you wouldn't be too stressed because you [would be] doing something, but instead you are just waiting, and waiting . . ."

Bobak Ferdowsi expressed similar sentiments and referred to NASA training sessions before the landing day, meant to prepare them for how to behave if the worst case happened: "We had had some briefings a week or so before about what to expect, how to deal with success or failure, some very real scenarios about how to deal with failure, since it's the more difficult of the two to handle." The agency needs people who can stay cool and focused even if things are not going well. "In your testing you realize that failure is not always immediately obvious, many of the scenarios don't involve realizing that everything broke and it's suddenly over. More often it's silence—we don't know what happened, it could still be alive, it could be dead, one of the radios could be dead and it could be a couple of days before it swaps to the other radio. Most of it is unknowns. So for us it was how to deal with this." Ferdowsi was also nervous and spent the night before landing in a sleeping bag in his JPL office.

The waiting was over on August 5, 2012. By midafternoon, the parking lots at JPL (which cover an entire hillside) were packed. The press was ushered to the media center, and by early evening most were set up and ready to report. Up the hill a few buildings, in the Space Flight Operations Facility or SFOF, the EDL team was assembled where they had been for many hours, preparing to wait out the atmospheric interface and entry. They would be passive observers as MSL made its way to the surface of Mars, alone and autonomous, fourteen minutes before anyone on Earth would receive a confirmation.

Seven minutes of terror: entry, descent, and landing had begun.

CHAPTER 17

ARRIVING

What follows is my account of how the evening played out. It's not necessary to follow this chapter with extreme precision to understand the rest of the mission—but the landing was so cool, and I enjoyed it so much, that I am sharing the nerdy, joyful minutiae. I wish everyone reading this book could have been there to share the moment—it was, for an old guy like me, not far behind the excitement of the Apollo 11 landing. You probably can recall your own comparative historical highlight. But most people experienced it somewhere other than JPL, so we'll relive it together.

As a "true believer," I headed up to JPL early. I also wanted to get a decent parking space, so I arrived before midday (parking is very limited at the lab's hillside location at the best of times). When I entered the media center, I saw a couple of familiar faces. As it turned out, a few old friends from my Griffith Observatory days were volunteering to help handle the onslaught of media needs for a large portion of the week leading up to landing. John Sepikas (that's *Dr.* John Sepikas to you) had gone on to become a math and astronomy professor at a local college, replacing the astronomy professor I had studied under and interned for there. The old chap's classes usually started out with about forty students ("Isn't this an astrology class??"—I knew then that they were in for an unpleasant surprise) and ended up with about five of us by the end of the grueling term, no exaggeration. Il Professore was tough. Sepikas has a far gentler touch and, from what I understand, works tirelessly to bring the subject to life for a new generation of students. Jim Somers was another Griffith expat who has completed a long career with Los Angeles County and in his semiretirement spends time working with Sepikas and his students, as well as staring at the sky through various telescopes. I also saw Bob Brooks that day, another Griffith pal, who has worked at JPL since the late 1970s on various programs, including Mars orbiters, and should one day at the very least have a Martian crater named after him. Finally, David Seidel, also of Griffith, has worked at the lab for decades designing and executing its excellent educational outreach programs. None of it got me any special treatment, but it was fun to catch up. Besides, I was there and at this point, many hours before landing, the most compelling thing in the media center (besides the displays of various JPL spacecraft from the past, and, well, the fact that it *is* JPL) was a series of video displays showing an ever-changing graphic of the current status of MSL. It was really interesting (JPL does provide the very coolest graphic interfaces), looking over the spacecraft's shoulder as it neared Mars. Current velocity, position, and other numeric values counted down on the margins of the display. It was a great way to see where things were and get a sense of the mission's progress.

Fig. 17.1. AS WE SAW IT: This frame is from the live video feed of the landing as sent out by NASA. Lower left readouts showed distance from the surface, velocity, and time until landing. Curiosity was fifty feet from touchdown at the time this was taken. *Image from NASA/JPL-Caltech.*

Over in the SFOF were the EDL team members and an ever-changing stream of folks involved in MSL but not critical to the landing. There was not a lot of spare room, so it was best to get your looks in early if you wanted to see what was going on in these last few hours before touchdown.

All over Southern California and across the country, public viewing stations and private viewing parties were getting up to speed. The Planetary Society, a non-profit started a few decades back by some of JPL's top retirees, was hosting a huge viewing event in downtown Pasadena where I had just completed signing a few books I'd sold at the adjoining Mars Society convention. Outside, a throng had gathered around a life-size Curiosity rover display that JPL had lent out for the day. It was great to see so many folks—the place was jammed—who were enthused about the landing. I hoped they would continue to be fascinated and perhaps pressure their elected representatives to vote through more money for the space program, but that's another conversation.

Even Times Square in New York got into the act, showing live coverage on massive video screens. While the crowd of about one thousand never approached New Year's Eve proportions, they were very enthusiastic. Other crowds gathered worldwide. JPL also sent out a continuous stream of tweets, cute first-person things that were signed by Curiosity itself. NASA's web servers were operating at full steam to keep up with viewing demand. The total US audience was estimated to be well over three million. So much for the public being "bored with science," as some detractors claim.

The hours counted down, and then . . . it was time. Entry interface (EI), the beginning of the "terror," was just after 7:00 p.m. local (Pacific) time. At this point, the radio delay between MSL and Earth was about fourteen minutes, so the craft was on its own, thinking autonomously. Everything was automatic and handled by the onboard computer. But that did not mean that the control-room personnel weren't watching with care and concern.

The rest of the times will be in minutes after EI (Please note: these times have been revised by NASA since touchdown; the times used here are correct, according to JPL, at the time of this writing).

ENTRY INTERFACE +/– TIME (IN MINUTES)

–10:00 minutes: The cruise stage, no longer needed, detached from the lander aeroshell.

BOBAK FERDOWSI: The charming young engineer who would forever become known as "Mohawk Guy" spoke to me at length long after the landing. *"Of course, we had had many dress rehearsals before that to make sure we knew what we were doing. So we just fell into the routine that we had used for most of cruise—there are procedures in front of you, polls are taken and you respond to the polls, see what the values are for the part of the software that you are looking at. In that way it is sort of calming, because it is a routine that you have been doing this whole time.*

"At this point it's all happening fourteen minutes ahead of what we are seeing, so, for instance, at this point we have just jettisoned the cruise stage, etcetera, so things are actually happening right now."

~~~~~~

−08:40: The spacecraft was de-spun from its long journey and stabilized.

−08:10: The entry guidance system was enabled; the first two of the tungsten weights were shed. This meant that the center of gravity of MSL was now offset from the center of pressure, meaning that it could glide off-center on its heat shield, generating lift.

00:00: Entry interface—MSL plunged into the Martian atmosphere.

~~~~~~

BOBAK FERDOWSI: *"A mix of things had been responsible for launch, cruise, and approach, to make sure it all worked together,"* he said, referring to all the parts of the flight right up to the time it entered the Martian atmosphere. *"There was some relief when we got to EDL because much of the stuff I was responsible for was over, and I thought 'Whew—at least that stuff worked!' You never want to be the one whose thing didn't work. I became very focused on landing at that point."*

He continued: *"It's like we had already taken an SAT [Scholastic Aptitude Test], and are waiting for the results—you've taken the SAT, you are done. Nobody tells you the score right then, you have to wait. We know that we have done everything that we can do, for many years now, so how did the exam go? So you're nervous, and something happening could change the outcome of your life. I don't have kids, so it's not like I can compare it to sending my kids off to college or something like that. Besides, at this point there is no way for the spacecraft* to phone home and ask us for help, like 'Hey, Dad, can you send more money?' It is just doing what it is going to do."

AL CHEN: A year after landing, Chen admitted in a press conference that he had experienced a heart-stopper early in the EDL sequence: *"I had two guidelines: don't say anything stupid and report everything you see. It's that second part I didn't do very well.*

"We had two types of telemetry coming back from the vehicle." At this point, this is the main data coming back from MSL. *"These 256 tones that Curiosity would send back tell us what was going on. That's all we had for about 14 or 15 minutes, from cruise stage separation until we got into the atmosphere. After that we started picking up data from Mars Odyssey.*

"Things were looking pretty good, we had ten minutes of pretty boring heartbeat tones, my blood pressure was up but it was just starting to calm down. . . ." Heartbeat tones tell the controllers the condition of the spacecraft.

"Then shortly after we hit the atmosphere we got a tone, called 'Data Catastrophic.' It's usually bad when you get these tones coming back that say catastrophic," he chuckled nervously, then continued: *"We had intended that tone to tell us that we were about to lose the vehicle. . . . I almost had a heart attack when I saw that. If it was true, it meant either that the vehicle was tumbling or it was not heat shield forward. . . .*

"So I thought about it for a second . . . I thought, if we are about to lose the vehicle, I'd better not say anything dumb, going back to guideline one. I thought, this should happen right when we hit the atmosphere, there should not be much error at that point. Do I really believe this? Is there more going on there than what I was thinking about? The computer was subtracting the readings from two sensors up there, trying to compare pressures, so I thought, let me just sit on this for a minute. . . .

"We were about to hit peak deceleration and peak heating, so it will become obvious if we are about to lose the vehicle in a minute or two . . . so I sat on it. Luckily it turns out that the instrument was working fine, we just had a calibration error. If I was not already sweating then, I sure was after that."

~~~~~~

+01.25: Peak heating was reached on the heat shield. It was pointed the right way.

+01:36: Peak deceleration occurred, straining the spacecraft at eleven times

Earth's gravity. MSL was still gliding on its heat shield, steering itself toward Gale Crater.

+04:00 to +04:10: The remaining weights were jettisoned, and MSL settled back to its natural balance, base centered downward.

+ 04:19: The 52-foot supersonic parachute was deployed at about 800 mph, or Mach 1.7. It supported 5,293 pounds of rover, aeroshell, and heat shield (minus the weight of the parachute itself).

<hr />

ROB MANNING: Manning agrees with Chen that testing parachutes is not like testing computer circuits or rockets. They are soft and somewhat unpredictable, and a real PITA. *"Parachutes are always a problem. The soft bits are the difficult part. Unlike metal for example—you can test metal, when you pull it, it will break under a given amount of force. You can't do that with fabric, though, it just doesn't cooperate. Fabric has a much more empirical trial-and-error aspect. So with the parachute it was a bit more trial and error. For MSL we tested the parachute at the wind tunnel up at Ames* [Research Center, a NASA field center in Northern California]. *One of the problems is when you mortar launch a parachute on Mars, the supersonic flow opens the parachute from the front to the back. On Earth it opens up from the back to the front, just because of the thickness of the air. For these tests we don't really care how it opens but that once it gets to its full state* [fully open] *that it doesn't break."*

<hr />

+04:39: The heat shield was dropped free. The MARDI camera began imaging the landscape below, catching a few shots of the heat shield as it departed.

<hr />

ADAM STELTZNER: *"We were now looking with the MARDI imager at the surface of Mars in a new way. We saw black sand dunes underneath us, and the red iron-rich soil that gives Mars its natural color. The attitude we have is governed by the trim angle of attack of the parachute."* As MSL descends, the parachute swings a bit, allowing the camera to do pan around. *"That gave us a sort of a tour of the neighborhood where we would be landing."*

AL CHEN: "*The landing radar was a concern here—with the enormous heat shield, and uncertainties about how it would wear, and the wear patterns affecting its aerodynamics, what would it do once set free?*" Would the radar lock on the falling heat shield instead of the ground? "*That was something we had to figure out—is the heat shield going to fall away fast enough so that all your radar won't get locked up on it?*" It fell away fine, and the radar saw the ground just as planned.

~~~~~~~

+04:44: The landing radar began feeding information about the remaining distance to the ground.

~~~~~~~

AL CHEN: Al relates that landing radar was a critical development for MSL. Such radar had been used before but needed to be far more accurate for this mission. "*If you don't know how fast you are going, certainly if your velocity is one meter per second, good luck landing at those speeds, because you have no idea where you are going. That was a big developmental hurdle. We accepted the fact that we would have to go off and develop radar that could give us very good velocity data down to the 10th of a meter per second level.*"

ROB MANNING: "*We tested the radar with a helicopter but we also tested it with an F-18 jet. We used the helicopter to simulate the descent stage with the rover dangling below and the radar on.*" It worked. "*The amazing thing is the only time we actually hung the rover from the descent stage was in the ceiling at JPL. We only did that once or twice because it was an expensive test, and you got everything you needed to know from one test. There is very little variability in the system.*"

ADAM STELTZNER: "*We were concerned about our landing radar. We had six beams on that radar, but we restricted ourselves to using only two when we separated the rover from the descent stage. Since we were only measuring two directions, but controlling for x, y, and z [three directions], we had to estimate one of them, and that was z. we were confident that we understood [the] surface gravity of Mars [and] that we could use that as an estimate for the z-axis.*" Z-axis in this case means down, the direction of travel.

"*But it turns out that there is a gravity anomaly at Gale Crater not present in any of our models. That meant that there was less gravity, by 400 micro g's, at the surface of the landing* site than we had anticipated. As soon as we separated and went to two beams, we started to slow down. We had set our throttle to handle what we had anticipated as the surface gravity, but the anomaly meant that we were slowing down.

"That was actually convenient because landing slower is usually thought of as better than landing faster, but it was sheer luck that the anomaly went the direction it did. If it had gone the other direction . . . I think we would have been fine, but every time something went wrong with the rover for the rest of the mission, the guys would turn around and say, 'It's because you EDL guys landed us too hard.' We got lucky. We need to remember that in the future."

~~~~~~

+05:00: At this time, Earth was out of radio contact so the "heartbeat" tones Al Chen referred to earlier could not be sent home to be heard fourteen minutes later, but this was expected. It would be sweaty-palms time nonetheless.

+06:00: The landing rockets were primed and made ready.

+06:16: Backshell separation—the protective aeroshell, as well as the parachute, were left behind to drift away, their work done. The landing rockets were ignited and throttled up to 20 percent. The rockets also guided MSL off to the side in what is called a "slew maneuver," maneuvering about one thousand feet to get it out of the way of the parachute and backshell, which continued to fall.

~~~~~~

ADAM STELTZNER: "This is the divert maneuver, we are getting out of the way of the parachute and backshell. We slew first to one side, then the other." The lander reoriented to move straight down at the completion of the maneuver.

~~~~~~

+06:18: The landing rockets were throttled up to full.

+06:40: Powered approach—all rockets were firing as MSL neared the surface.

+06:43: MSL decelerated further to lose speed.

+06:50: Four of the landing rockets throttled down, and the remaining four were shut off.

+06:53: The rover began to rappel down from the rocket pack, descending on nylon cords. The front wheels were released to fall into position for landing.

+07:01: The "bogie" was released, completing the wheel-suspension-system preparation to act as landing gear.

~~~~~~

ADAM STELTZNER: *"We liked that we were using the wheels and suspension system, which had been designed from the beginning to conform to uncertain terrain. Slope-related disturbances would not affect the vehicle. Even if the rover could not rove, due to the steepness of the slope, the landing would be successful."*

~~~~~~

+07:02: The computer enabled "touchdown logic," the software that was needed to complete landing.

+07:11: Touchdown on Mars.

+07:12: The rover's onboard computer confirmed that all wheels were on the ground and stable.

~~~~~~

ADAM STELTZNER: *"It was decided not to measure the landing event, but to measure the state of the vehicle."* What's the difference? *"We wanted to avoid the fate of the Mars Polar Lander, which had a false indication, noise in the system, which it interpreted as a landing, shut off the rocket motor and fell the last 80 meters."*

~~~~~~

+07:12: The flyaway command was sent to the rocket pack, telling it to go crash on its own somewhere else, at least 2,100 feet away. The "bridle," or nylon cords along with the wires leading up to and throttling the rocket pack, were cut.

〜〜〜

ROB MANNING: *"When the wheels made contact, the rocket pack continued coming straight down, slowly, so it never hovered [prior to touchdown]. When the rover sensed ground contact, the computer realized that the rockets were using less power to stay up in the air. The computer then says 'I must be on the ground.' So it sent the command up to the rocket pack to say 'Stop and hover,' so now the descent stage begins to hover. It slowed to a stop and finally the computer cut the cables. Before it cut the last electrical cable, it said, 'Okay were done, go ahead and fly away!' in one second the descent stage replied 'Right—copy that.'"* He said this last part in a cute British accent. I wondered when the rover started talking like a Spice Girl, but didn't ask. The important thing was that it had landed safe and sound.

〜〜〜

About fourteen minutes later, on Earth, the whole drama was played out for humans to hear. A bit over seven minutes after this delayed transmission had indicated entry interface, Al Chen confirmed touchdown. The control room was jubilant, as were three million viewers.

Ferdowsi summarized his feelings after landing, which doubtless applied to many in the room that night: "I was super happy once we had landed——it was exciting to share it with everyone else, you know how hard everyone has worked up to that point. What surprised me was how leading up to that I had felt like I was getting tired, you've worked these long hours and done all these things. But the moment we landed my first thought was: 'Let's do this again! I'm ready, let's go and launch another rover now!'"

A very tired Adam Steltzner wrapped up the night with a press-conference statement that further inspired: "Great things take many people working together to make them happen. That is one of the fantastic parts of human existence. I'll be forever satisfied if this is the greatest thing that I have ever done. . . . This nation is a truly great representation of a piece of humanity that reaches out and explores and conquers and engineers. We are toolmakers, agriculturalists, and pioneers . . . and that is reflected in the activities and actions and results of tonight."

Well said.

Fig. 17.2. JOY: After the landing, the generally restrained engineers cheered and wept with joy—a decade of work had paid off brilliantly. Adam Steltzner embraces Al Chen in lower left. *Image from NASA/JPL-Caltech.*

CHAPTER 18

WEIRD WATCHES: LIVING ON MARS TIME

In a swath of suburban bedroom communities hugging the foothills of northeastern Los Angeles, people began to notice odd things. These normally quiet and pedestrian communities—La Crescenta, La Canada, Flintridge, and Pasadena, are not places where people generally act out much, and disruptions of routine are noticed. These are towns where teenagers' parties are often shut down by the cops at 9:00 p.m. But what the neighbors were seeing looked like a low-budget slasher movie, where the neighbors just aren't quite right and you probably should begin to worry and round up the kids before they are trapped and cooked. People who normally worked a pretty steady 8:00-to-6:00 schedule were coming and going at odd hours. When home, they were drawing heavy black shades across their windows. Sometimes entire families, even young children, were seen having barbeques complete with ribs and beer (for the adults, of course) at 7:00 a.m. Something was not right. It was all a bit *The Hills Have Eyes*.

But there was at least one non-JPL local who knew what was up. Garo Anserlian, a jeweler in nearby Montrose, was helping these folks to maintain their bizarre new schedules. For besides being a jeweler, he was also a part of a rapidly vanishing breed, a watchmaker. And he was creating some very odd timepieces for a number of folks at the lab.

Starting with Pathfinder, it became clear that early-stage operations of landed rovers worked best if the engineers and scientists responsible for wringing the most out of the machines lived on a Martian schedule. Mars days, or sols, are much like Earth days, lasting 24 hours . . . and 40 minutes. So if you spent three months on Mars Time, as they had on the Mars Exploration Rovers, your schedule shifted pretty quickly. You lost about five hours per week.

Keeping track of this became a major pain in the neck, so a couple of engineers on the MER project had decided they wanted watches devoted to keeping the oddball time properly. But what digital chip maker would design something for the few dozen people who might spend a couple hundred dollars for such a timepiece? The answer was *none*.

The engineers approached Anserlian, as he was one of the few locals who could even fix a watch, much less make one. He said he'd check it out. He consulted other watchmakers, including some acquaintances in Switzerland, and they said forget it, it's too hard, don't get mixed up in it. Fortunately, he was an inveterate tinkerer and ignored them, spurred on by the challenge.

Starting with a mechanical watch, he added weights to the balance mechanism and adjusted the spring, and soon he had a workable 24-hour, 40-minute watch. The JPL'ers gobbled them up, and many were still in use when Curiosity landed. New ones were still available at a price. Many of the watches even had red Mars logos on the face.

None of the neighbors thought to ask Anserlian what was going on, but they should have. He would have told them about Mars Time with great pride.

But buying a Mars watch was the easy part. Shifting one's circadian rhythms— even at the slow rate of forty minutes per day—threw one's system out of whack. Some tried pharmaceuticals, most just dealt with it. None were unaffected.

And it's not as if this was a simple eight-hour workday. Most people on the mission labored from twelve to fourteen hours because that is what it took. That alone can disrupt a regular schedule—adding almost an hour to each day made it even tougher.

Most JPL'ers I spoke to said, however, that very strong bonds formed during Mars Time. The LA-based people already knew each other pretty well, having for the most part worked together for years, but the others from faraway cities and other countries were brought to JPL for the three-month duration of primary operations in Mars Time. So, although the days were drifting continuously, everyone was working together on long but synchronized shifts.

Rebecca Williams, a participating scientist on MSL, now works on the mission out of her home in Wisconsin. So, in addition to the Mars Time shift, she had the time zone change to deal with as well. Or perhaps that already gave her an edge when she came to Pasadena? She spoke to me from her chilly basement office just outside of Madison.

Fig. 18.1. MARS WATCH: One of Garo Anserlian's Mars timepieces. They come in many styles, but all have in common the 24-hour, 40-minute Martian sol time measurement. *Image from NASA/JPL-Caltech.*

"In the beginning there's a lot of engineering check outs, but it was really a fortuitous thing that we all got to live and work together for three months. You develop important relationships, and it makes it much easier to do the telecommunications that we do now on a daily basis. You know whom to contact and you've developed that relationship so you feel more comfortable. It also increases the efficiency of doing it, which was really amazing. Of course, it was already a bonding experience to be exploring this amazing area on Mars, but doing it on Mars Time so you're all jet lagged . . . that's when you really get to know people."

It sounded to me as if it had been slightly easier for her to adjust, as she was already shifting her time to accommodate the West Coast clock.

"Those three months were among the best I had ever experienced. It's kind of like the most idealized summer-camp experience you've ever had. It's as if you're synched with the people you most admire and you're all experiencing something amazing together for the first time." Her voice almost quavers with rich memories of the early days of the mission. "One of my favorite experiences, and I still can't get over this, was when we had data coming down and everybody would gather around and watch as the new images flickered up the screen. You would look at a rock and you think you understood it, then the person next to you had a completely different interpretation. So you are going back to your first principles as to why you think you are interpreting a rock a certain way. We all saw them simultaneously for the first time. It was a lot of fun."

That does sound like a blast. I asked others to see how their experiences compared.

Lauren DeFlores had worked on the MER mission, then the Mars Phoenix lander in 2008, so she was by now an old hand at Mars Time. But Curiosity threw her a new challenge—or perhaps it's more precise to say that the mission's timing did. She was an integration engineer for the ChemCam and DAN instruments, but she was also soon to have her third child.

"Mars Time on MSL, the first ninety sols, was also my first trimester of pregnancy. I actually think it worked out really well because no one could tell how tired I was because *everyone* was tired." She smiled. "I think one of the greatest things is that our project management does do is every time there's some sort of large event they think of our families first. That provides unconditional support, dealing with people who are working with ungodly hours. So during that whole ninety sols I was working probably ten to twelve hours a day, or per sol, only sleeping for a few hours, and coming back for another shift. It was pretty rough in the first months, and then it got a little better. I was pregnant, but my husband took care of the kids."

Good man, he. I began to think that this Mars Time thing was pretty interesting, so I thought, why not try it? As a writer, I work bizarre hours anyway, so I gave it a shot for a week. At first I did not notice much. And since my hours are normally all over the place, it was not a big deal. So I extended it a few days . . . and then it begins to really catch up with me. By day 6, my cat was looking like me as if to say, "What the heck are you doing in my part of the house at this hour?" and my son simply slept through it. I gave up. Somehow, without the unifying experience of being on a mission, it was just not the same writing about it and being sleep deprived.

While *my* son may not have been interested, participants in the mission with younger children found it sometimes made quite an impact. David Oh, a flight director for the mission whom I spoke with as Mars Time came to a close, took an unusual step. "For the first month after landing, my whole family joined me on Mars Time. And we jumped a time zone per day, every day, for thirty days, going all the way around the clock. We got to explore Mars at JPL and Los Angeles at night. It was a great adventure for my whole family." His kids seemed grateful that they found an International House of Pancakes that was open twenty-four hours.

At the end of the three months, shifts adjusted to normal working hours, and many nonlocal team members went home, adding jet lag to the wacko schedule change. For some it was three hours to the East Coast, to others it was ten or twenty as they headed off to Europe or Asia. But to a person they agreed: Mars Time, grueling though it is, is a wonderful and unifying experience. Watchmaker Garo Anserlian would surely agree. After a surge during the early days of Curiosity, he still sells the occasional Mars watch to anyone willing to pony up the few hundred dollars it costs to make one. When I have a few hundred extra dollars, I just may treat myself to one.

CHAPTER 19

THE ORBITERS: HELP FROM ABOVE

Exploring Mars is a team effort. From a human-resources point of view, it has been such since the 1960s. Prior to that time, Mars was a telescopic object. As we know, astronomers like Schiaparelli and Lowell spent hundreds of hours at the eyepiece, drawing what they saw. This was by its very nature a solo or small-group endeavor.

When NASA and the Soviet Union's space-exploration efforts got underway in the late 1950s, large teams were needed to accomplish the gargantuan tasks placed before them. By the time of the United States' Mariner missions to Mars, small groups of senior scientists worked with larger groups of subordinates—both science people and engineers—to accomplish these large projects.

Today, partnerships surrounding Mars have taken new forms. The intricate operational ballet between the rovers on the surface, the orbiters circling the planet, and even Earth-based observations of Mars, require great precision in planning and execution to be successful. These overlapping data streams, when properly coordinated, have provided ever more detailed and dense sets of information about Mars—its surface, atmosphere, geological composition and characteristics, radiation environment, and much more. Without this suite of robotic explorers and people working in concert, missions like MSL would not be possible.

In the 1990s, in the same time frame as Pathfinder, came the largest advance in orbital imaging in twenty years, the Mars Global Surveyor. This is the mission that set Mike Malin and Ken Edgett onto the path toward realizing that there had been sedimentation processes taking place on Mars. It was the first real jump in understanding the Martian surface since the Viking orbiters mapped the planet in the 1970s, which in turn set the stage for the MER and MSL missions. And, while it was generally accepted since the early 1970s that water must have had some role in shaping the tortured landscape the orbiters saw below them, the MGS mission was the first time the scientists began to understand the true nature of, and role of, water in the grand story of Mars.

Fig. 19.1. EYES ON MARS: The Mars Global Surveyor spacecraft arrived at Mars in 1997 and operated for nearly a decade. It provided the first high-resolution look at Mars and the first orbital images since Viking in the 1970s. The tube at the bottom of the spacecraft is not a rocket, it is Mike Malin's camera. *Image from NASA/ JPL-Caltech.*

Then the current generation of orbiters were sent out to Mars—the Mars Odyssey orbiter in 2001 and the Mars Reconnaissance Orbiter in 2006. Both were eventually tasked as relay stations for the MER and MSL rovers, handling the retransmission of the copious data that comes back from the ground-based machines. It has been a huge advantage over the line-of-sight, ground-based messages that Pathfinder was restricted to, opening up the bandwidth tremendously. And as we have seen, bandwidth is a huge enabler of—or limitation to—science performed on the surface of Mars.

Also working above Mars is the European Space Agency's Mars Express (ME) orbiter, which had provided its own trove of visual and other enticements about Mars starting in 2004. One key observation of ME was the apparent existence of measureable amounts of methane in the Martian atmosphere, which can be an indication of biological activity. Since methane does not last long in the Martian air, the reasoning was that it must be continually replenished to be observable. Curiosity has tried to sniff for methane on a number of occasions since touchdown but has failed to detect any, so the jury is out on this one . . . it might be that the ME indications were erroneous, or perhaps that there is some source of the methane that has nothing to do with processes taking place in and around Gale Crater where Curiosity's measurements were taken. It's nice to imagine a huge subterranean grotto filled with Mars microbes—or Mars cows, six-legged giraffes, or *something*—generating the gas somewhere near the poles where it would be harder to detect. But that's just my own wishful thinking. I'll always be a bit of a Lowellian optimist. I love science and adore the Mars exploration program, but it would have been keen (to use the vernacular of the day) if Mariner 4 had spotted some canals and a few oceans or forests. Or a herd of a thousand eight-legged thoats (Edgar Rice Burroughs's imaginary Martian steed), moving in unison. But once again, I digress. Back to reality.

Before the orbiters would give a helping hand with rover data, they would first assist in defining a place to land. By the time the MER mission plans were being finalized, with guidance from Pathfinder and the Mars Global Surveyor data, water was becoming the primary focus—some might say obsession—of the Mars program. There is lots of other great stuff to study, but water-based processes provide tremendous information on the weathering, geological transformation, and landscaping of Mars, and of course is critical to life as we understand it.

Central to the observation of watery processes on Mars is how it has affected the terrain, exactly what hand it had in the evolution of the Martian environment, and, hopefully, an understanding of when those processes occurred. We've already learned how Mike Malin's observations and hypotheses about water—as seen from orbit—affected some of the landing-site candidates for Curiosity's mission. It's also worth understanding how the orbiters assist in such research. And who better to tell us than Ken Edgett—Mike Malin's trusted associate, the man who spent the better part of a decade working ten- to fourteen-hour days looking at well over a quarter million (yes, over 250,000) images of Mars taken from orbit, and the same guy who co-led our grueling trip to Death Valley.

Edgett is a consummate space scientist and geologist, which is to say that he *lives* the science. In his spare time—what little there is—he writes children's books about science as well as adult-level science fiction, both of which have been published. Prior to joining Malin Space Science Systems (MSSS), he developed and led a science-education program for kids called Mars K–12 at Arizona State University, where he received his doctorate. He has also worked on children's science television programs, and much more. If anyone is committed to bringing science to young people, this guy is. You can *feel* his passion for Mars. He practically crackles with excitement, and he is one of the few scientists of his stature whom I have met who will willingly allow emotion to color his expression; not in a gooey, join-hands-and-sing-Kumbayah sort of way, but in confessing that remembering certain moments, like when he realized the true nature of a site he was researching on Mars, or a moment of discovery with MER or MSL, choked him up and still can. Hell, some of those same moments choked *me* up and I had no involvement in the program. Had I been in his shoes, directly responsible for some of these discoveries, I probably would have fainted dead away.

I'm going to let him tell his story because at its best, science is the unraveling of a grand mystery, and who doesn't love a good mystery? What is important is how it is told. I spoke to Edgett from his office at MSSS. We began with his work on the Mars Global Surveyor mission, for which MSSS built the high-resolution camera.

"Every day, starting in 1998, I was in here staring at maps of Mars, looking at where the spacecraft's orbit would go and try to decide what to take pictures of, along with Mike and his team. But I did a lot of it. As we gathered this data, we started to look at what are the most important things that we were finding. There was just so much to discuss. You're peeling the skin off of the cream that rises to the top. There was so much good stuff, we didn't even get to do the cream, so to speak!" This was when I began to understand Ken's passion for Mars. You could hear it in his voice, the way some people talk about their most cherished memories. He then spoke of some of the same pivotal scientific papers (the ones Grotzinger had mentioned to me) as being critical to understanding sedimentation on Mars and being responsible for revising the process of choosing landing sites. "One of the important papers was in 2000 . . . what we both [he and Malin] thought of as the most important from that whole [MGS camera] experiment was the sedimentary-rock observations. Before we did that work in 2000, when people talked about Mars they did not put the words *sedimentary* and *rocks* together in the same phrase or the same sentence. It just wasn't something that people were thinking about.

Fig. 19.2. KEN EDGETT PONDERS: Ken Edgett listens to a question during a press conference at JPL. His folksy demeanor and passion for K–12 education has encouraged many young people to become involved with the sciences. *Image from NASA/JPL-Caltech.*

"I could show you a couple of abstracts [for scientific papers] where people speculated that certain Martian meteorites might have been sedimentary rocks, but that was not the thinking of the mainstream Mars science community. People weren't thinking about the presence of rocks on Mars formed from sediment. We knew Mars had sediments because it had sand dunes that have formed by wind—that's a sediment, but they are modern dunes [that is, soft sand and not sedimentary *rock*]. We knew that the polar caps have layered stuff in them as well. And the thinking was [that] this was a mix of dust that had settled from the atmosphere, which would form a sediment, and ice, which froze in the atmosphere. We knew about that since Mariner 6 and 7. From Mariner 9 we also knew that there were exposures of layered material in places like Valles Marineris [the largest valley on Mars].

"The key item here is that while these areas were recognized as containing layered deposits of some kind, they were not thought of being like earthly, old cemented sedimentary rock. It seems odd, looking back, but it was just not the way people looked at it. They weren't discussed as being *units* of rock, and the distinction of that mind-set is they might not have been, they might have been piles of loose, unconsolidated layers of dust or ash or something."

But when someone spends eight years looking at hundreds of thousands of photos of Mars, images of increasingly higher resolution and clarity, something seems bound to shift. "Back in 1999 and 2000 we were seeing all these layered rocks. And we could make the case that these were *rocks*. They were hard. They held cliffs. The cliff would tell you something about the hardness of the material. Craters could be preserved in them. So we told that story in the paper that Grotzinger was talking about. That was a tipping point . . . that Mars is different—there are sedimentary *rocks*."

Then, among the literally dozens of other papers Edgett worked on during those years, came the 2003 discussion of sedimentation on Mars in a river-delta region near Eberswalde Crater, an area that became a top contender for an MSL landing. "To Malin and myself that area was the smoking gun that some of these sedimentary rocks really were deposited by water. When we wrote that paper in 2000, we talked about bodies of water because a lot of these are in craters, and craters could be a basin where you might have water but couldn't really demonstrate that water *did* anything." They had to consider the alternatives to water-deposited sediment to cover all the angles. "We also talked about ways to [create the landform] purely by atmosphere-blown sediment and things like that." That is, windborne sand and dust as opposed to water-moved sediments.

They were being cautious, but what they were seeing seemed to lead in only one direction: "One of the things we saw were places where you have repeated layers of similar thickness repeating over and over again." Rocky strata. "Our experience told us how you can do that with the right kind of fluid properties." Again, earthly analogues were supplying information that could be applied to what they saw on Mars, and the conclusion led to water.

In 2003 they were able to confirm their suspicions. "We saw a delta in Eberswalde Crater. That feature is sedimentary rock. It is hard, it has eroded, and the erosion enhanced the appearance of the feature. Where there were stream channels in the delta, now those channels are ridges, they are raised above the surroundings because now they are coarser or more cemented than the surrounding material. That was an indisputable smoking gun that there was a river, a delta, and it went into perhaps a body of water." It was conclusive evidence of water-deposited, hard-rock strata. And a smoking gun is always welcome in a good mystery tale.

"These were very ancient sediments that record what kind of environment it was back whenever they were deposited. So then when you fast-forward to the question of why [we chose] Gale Crater and Mount Sharp, the answer is because you have a five-kilometer-high pile of these sedimentary records, which are records of different environments over some period of time. Of course we haven't gotten to Mount Sharp yet and we will never actually do the whole five-kilometer thickness of it, but you have got to think big!" The last statement is said with a flourish, and it makes sense, coming from him. Both these men, Malin and Edgett, had to be thinking big to make the claims they made when they made them. A lot of scientists in a similar situation might have sat on the idea for much longer or waited for even more (possibly redundant) evidence. It can be dangerous to swim upstream too soon in scientific circles.

Now, when most of us see something compelling and convincing, we form an opinion. But science is a process of forming an opinion based on observed evidence (your hypothesis), then working like mad to prove that you are wrong before you proclaim that you might be right. Edgett discovered an irony—the more images he stared at, the less certain about some things he was. "What happens is that when I see as much data as I have, I find that it makes me more reluctant to really jump in and try to interpret any of it! In some ways it makes me a better scientist, and in some ways it makes me a worse scientist. The way it makes me a worse scientist is that I am not publishing, and you know it is a publish-or-perish world." Huh. You would think that a guy who published six major papers in 2000 alone might not have to worry, but what do I know.

"The way it may make me a better scientist is that I've seen a lot, I've observed a lot, and I still feel like I'm still doing that now with the data from the Mars Global Surveyor and Curiosity. A lot of what I have observed is still in my head being processed." He made an analogy of how young children who learn multiple languages from the start sometimes take longer to speak any of them. "I've been doing this for going on sixteen years now, and I feel as if I'm still learning the language. I'm not ready to speak, I'm just learning it. What happened in my case, because I look at all this data and I see all these things, you think you know what's happening on Mars. Then Mars throws something new at you, and you suddenly realize what you thought you knew is garbage and you've got to think it through again. And I always get knocked backwards by that process. I should point out this doesn't stop everyone in my field right? The science goes forward, and people [are] finding great things and writing papers about them and announcing discoveries all the time. But I'm always a little more worried and cautious because I've seen too much."

He is being kind to many of his fellow planetary geologists. There are people who have seen far less who have advanced theories based on that slimmer volume of data. But when you see so much (over a quarter million images!), it seems that the odd result is that there are more specific instances in your mind that can cause doubt. It's scientific humility created by observation and introspection . . . and the oft-mentioned concept that Mars will always, always try to screw with you. Maybe the Greeks were right about Ares . . . he was an irresponsible, irritating oaf as a god, and still can be as a planet.

This brings us to an interesting junction: we've seen how looking at Mars from orbit helped us to understand that there were sedimentary processes taking place, and by extension, in the case of MSL, this resulted not only in a well-crafted selection of a landing site, but it is what brought John Grotzinger to the table in the first place.

What happens when orbital observation meets ground truth?

"It's really interesting for me now having the MAHLI camera down on the surface. Now I'm taking what I've learned from orbit and I'm going all the way down to things smaller than a millimeter." That's one heck of a change in scale. "I'm trying to wire those things back up to what I see from orbit. It's kind of cool."

What will Malin and Edgett make by combining their massive catalog of orbital images from what they learn on the ground? You'll just have to wait and see.

CHAPTER 20

FIRST MOVES

C uriosity was on Martian soil. Sky crane had worked. The descent stage had flown off, crashing as planned some distance away—it would be imaged, and its point of impact refined, later. The same was true of the heat shield and even the tungsten weights that had been ejected from MSL on the way in. All these impacts were important to track, if possible. The images of the shape, depth, and even direction of the small craters they made—from which soil properties could be established—were all important. Also important was an evaluation of the images from the MARDI camera during the final phases of descent. JPL would set people to the tasks even as they commissioned the rover.

May I say it again? Nothing is wasted in space exploration.

But first and foremost, they needed to make sure that Curiosity was healthy, stable, and ready to go to work. The official landing time was 10:32 p.m., Pacific Daylight Time on August 5, 2012. The overall mission was still formally called the Mars Science Laboratory, but from now on, most people would just refer to it as Curiosity. That is the name of the machine now roving the Martian surface.

In those early hours at Gale Crater, the engineers were busy making sure that the onboard systems were healthy and waking up properly. Meanwhile, Mike Malin and the imaging team were working feverishly to bring back the first images of the newest part of Mars to be explored. The very first picture, taken by one of the low-resolution Hazcams aboard Curiosity, showed the shadow of the rover and some terrain in the distance. The image was canted to one side and somewhat occluded by the dirty lens cap (which would soon be opened), but it was proof that Curiosity was down, safe, and operational.

At the same time, still frames from the MARDI descent camera were being stitched together to create a low-resolution depiction of the downward-looking image stream downlinked in the final phases of landing. A high-resolution version would follow some days later. It was spectacular.

Two hours after landing, Curiosity sent back another Hazcam image, this time with the dust cap removed and with much better clarity. Though it was still distorted and low-fidelity (only 512-by-512 pixels, or about a quarter of an HD television frame), it was not long before Grotzinger was beginning to pose questions, just as a good chief scientist should: "Curiosity's landing site is beginning to come into focus," he said in a press conference. "In the image, we are looking to the northwest. What you see on the horizon is the rim of Gale Crater. In the foreground, you can see a gravel field. The question is, where does this gravel come from? It is the first of what will be many scientific questions to come from our new home on Mars."

Fig. 20.1. EARLY IN: One of the first images sent down by Curiosity. This is pointing toward Mount Sharp and was taken by one of the Hazcams. *Image from NASA/ JPL-Caltech.*

Curiosity was not working alone on Mars. NASA's orbiters, Mars Odyssey and Mars Reconnaissance Orbiter, were providing images that pinpointed the landing site and the condition of the rover using their high-resolution cameras. The data were of particular interest to the JPL'ers who had entered a game John Grotzinger had put together called "Landing Site Bingo." Team members had earlier been invited to go to a wall-sized printout of the landing ellipse and pick a spot, tagging it with their name. Attempting to factor in wind speeds, atmospheric density, and uncertainties about the exact behavior of the parachute all added to the fun. When the orbital pictures were in, the winner of the competition was our very own Rob Manning. Upon hearing the results, Grotzinger quipped, "No doubt about it . . . it was fixed." As it turned out, they had landed less than two miles from their preferred landing site. Considering the vast distances, the autonomous operations of the lander, and the vagaries of the atmospheric conditions, it was an amazing accomplishment.

A low-resolution series of images was returned from the main cameras atop the mast and quickly built into a panoramic, providing a good view of where the rover was in relation to ground features and Mount Sharp. High-resolution images followed.

These pictures revealed that pebbles had been thrown up on the rover's top deck during landing. This accounted for why one of the two wind sensors was not operational—a rock kicked up during landing by the rocket motors nicked one. It was an inconvenience, but they had a workaround in short order. A side benefit of taking test images with the Hazcams was to ensure that all the transparent, plastic dust covers had swung free—the cameras would still be usable even with the filter hung up, but the images would have been degraded, just as the very first image had been.

At the same time, "commissioning" continued. This is the process of waking and checking the various electronics and instruments on the rover to make sure they had all survived the rough journey to Mars. Everything (with the exception of the one damaged wind sensor) was operating within specs.

As the imaging team began to build a three-dimensional ground-level map of the surrounding terrain, the engineers turned on and checked the backup computer.

Each day during this commissioning phase, the ground team "awoke" the rover by playing a piece of music themed to a day on Mars. Numbers ranging from "Good Morning" from the musical Singin' in the Rain to Beatles tunes were played. The rover certainly didn't care, as it was perfectly capable of waking itself from the cold Martian nights. It was, however, a nice morale booster for the increasingly jet-lagged operations team working on Mars Time.

Fig. 20.2. EARLY PANO: An early black-and-white panoramic image of Mount Sharp. As time allowed in the first few days, more color images were sent back and stitched into smooth panoramics. *Image from NASA/JPL-Caltech/MSSS.*

On the second sol, the orbiters had pinpointed where the last six tungsten weights had impacted. This gave good data on ground conditions at these spots, derived from the amount and type of soil disturbance. They landed in a pattern about 7.5 miles from the rover.

Instrument checkout continued even as the Mastcam took a high-resolution 360-degree panoramic of the surrounding terrain. The RAD (radiation detector) took readings of surface radiation levels.

Working with the new surface imagery and orbiter photos, rover planners were able to incorporate both overhead and ground-level images to begin building extremely accurate virtual maps of the surrounding terrain. This allowed them to see features and obstacles beyond the rover's line of sight, which was a valuable planning aid. This had been pioneered with the MER rovers but was now available from the very beginning of the ground exploration. These cameras, working in concert, would allow for safer drives and gave the onboard computer "awareness" of the terrain surrounding the rover. The autonomous driving it would soon be doing would depend on these maps.

A major event was scheduled for five sols after arrival. The onboard computer had been loaded with software for space travel and landing—information that was no longer useful or needed. The programmers had a new batch of software that needed to be installed. They had uploaded this programming while the spacecraft was still in transit from Earth; now it was time to move it from storage to the main computers. It was to be a three-sol process that they referred to as a "brain transplant."

Just to be on the safe side, the flight software had included basic operations for the rover—driving and so forth—just in case there was a problem loading the more complex and capable upgrade. They could still make progress even as issues were worked out, but the upgrade was needed to enhance autonomous driving and operation of the robotic arm.

Other operations were suspended as the "transplant" took place, but the assembling of images from the Mastcam continued.

At this point, JPL was still holding on-site press conferences. These would soon become teleconferences—it's likely more cost-effective and a better use of scant resources than continuing to host the ever-shrinking number of news-media personnel that are closely following the mission. The major news outlets have fairly short attention spans for such things—the drama of landing was over, the first surface images were back. Now it was just more of the same, right? Well, not exactly. Each day revealed new and amazing surface images both from the rover and the orbiters. But that is the food of specialized media, such as Space.com and the Planetary Society (among others, including books by this space geek . . .), so those of us working for *those* outlets were going to get as much on-site time as we could.

I was there for the release of that first high-resolution panoramic shot. The images had to come down, one by one, and then be assembled into a wide strip of the horizon with color correction applied. The one we saw first, in true "Martian color," was the expected dull, ruddy hue of Mars. It looks like a very smoggy day or, if you happen to have lived in Southern California, much like a day when the summer wildfires are burning in the foothills—the landscapes and sky glow a dull brownish red. The white-balanced version shifts the image toward the blue, simulating light on Earth, and the geologists can see the rocks as they are used to seeing them (for example, during terminally strenuous field trips to Death Valley . . . not that I'm fixated on that or anything). It was the latter image that caught my attention. . . .

To my everlasting shame, my first thought was that it looked like a gravel-strewn Walmart parking lot, complete with the San Gabriel Mountains in the distance (had the Walmart itself been visible, the day would have turned out very differently). Of course, it was Gale Crater with Mount Sharp in the background, but the foreground was so flat and sharp that it was immediately unremarkable. Then the reasoning brain caught up with the primitive one, and the wonder of it swam into view. It was, to apply an overused adjective, spectacular.

It is interesting, however, that for a person who came of age during the "golden era" of space exploration—Apollo to the moon, the Vikings to Mars, and the Voyagers to the outer planets—it looked almost *too* good. The people at the lab are so good at their jobs, and Mike Malin's cameras are so good at *their* jobs, that it made it look almost too easy. Of course, this is silly and counterproductive, spawned by memories of watching NASA struggle with the challenges of those early pioneering days.

One thing that is remarkable with the newer cameras, though, is the color. The Viking images had been orange, so saturated and deep that they looked incredibly exotic. What we were seeing now looked much more like a day out in the redder regions of the national parks in Utah—crisp, clear, and a desaturated, dull brick red. The sense of wonder is now more subtle.

Then when you spent a bit more time looking at Mount Sharp in the distance, and realized that within a couple of years the rover would be *climbing* in those foothills, the sense of awe came back in a blast. This thing is actually going to dive to, and in some cases *through*, those places, and we can see them like they are just across the street.

There was also an image of the crater wall opposite Mount Sharp, the rim of Gale Crater. It showed what was clearly the result of fluvial processes—a river or stream entering the crater from the surface above. It was the first ground-based observation of such a feature at that scale, and it drew deep sighs from those who knew what they were looking at.

Closer to the rover, the cameras spotted the areas nearby where the descent stage's rockets had briefly scoured the surface. The gouging was not deep, but the rockets had been sufficiently forceful to blow away the loose surface soil. What remained was the darker, hard surface underneath and rocks embedded in it. This was a free bonus provided to the geologists by sky crane, and it would be looked at in more detail later.

By the end of sol 5, the software upgrade was successfully crossing into the main memory and was tested.

The next day, a priority call was received from Air Force One. The president wanted to say thanks to the Curiosity team. The lab's director, Charles Elachi, fielded the call and responded that he hoped the mission would inspire the many young people who were watching the landing and were tracking the progress of the rover from classrooms all over the country. Of course, the president did not miss an opportunity to give a shout-out to Ferdowsi: "NASA has come a long way from the white-shirt, black dark-rimmed glasses and the pocket protectors . . . you guys are a little cooler than you used to be." He continued by saying that he had "thought about getting a Mohawk myself, but my team keeps discouraging me." The president closed with, "You guys should be remarkably proud. Really what makes us best as a species is this curiosity we have—this yearning to discover and know more and push the boundaries of knowledge."

The call was piped throughout JPL and across NASA's television network, and then it was picked up by major news outlets. Despite the strained budget, despite the trimmed programs, a lot of smiles were seen around NASA that day.

On sol 9, JPL announced that the new programming had loaded and tested out properly. The rover's brain was digesting the new software and executing the commands as expected, which was a huge relief. While nobody expected any trouble, programming computers from over 150 million miles away is anything but routine and kept people's attention. The relief was palpable.

CHAPTER 21

MEETINGS

In my years as an academic employee—both on the administrative and professorial side—I, like so many before me, grew to have a genuine loathing of meetings. Academic meetings have an amazing tendency to crawl on for hours and achieve astonishingly little. A certain form of order—roughly Robert's Rules—are followed, which by turns give structure and destroy enthusiasm. The most challenging part for me was that once on the dreaded academic committee, one could attend monthly or bimonthly meetings for a year and sometimes accomplish almost nothing. Of course, not all committees are like that—maybe even most are not—but the few I was pulled into were. I commented on this once, after enduring a year of long monthly meetings on the subject of academic retention—keeping students through until graduation. I was returning to life as a television producer, and I thought, as a closing flourish, I'd make an observation. At my final meeting, I was asked my opinion on the proceedings. "I've never spent so much time accomplishing so little," I said with a straight face, "And am astonished that we are discussing, meeting after meeting, the same things we began with." A few recoiled in shock (you are allowed to think these things, but never to *say* them), others offered a knowing snicker, most just stared at the center of the broad Formica expanse that separated us. Needless to say, the chair was displeased and sputtered a bit.

I can't blame those who were displeased—it was not good academic form—but it was satisfying nonetheless.

So you can see that I have a distaste for meetings. But on sol 506 of the Curiosity mission, my faith in meetings was restored somewhat (yes, we are jumping ahead here, but only for this chapter. The activities are sufficiently similar on most days—in very general terms—that it's good to get an idea of how the teams work before proceeding).

When the chance came to attend a planning meeting at JPL for the next two days of rover-driving activity, I was not sure what to expect. The tireless Guy Webster, PR lead for Mars Exploration at JPL, called me to let me know I had been granted access to a strategic planning session at the lab for Curiosity. Being the gentle soul that I usually am, I put my previous experiences in a bottle and shelved them in order to approach this effort with a fresh attitude. But it never takes long to uncork that container should conditions warrant. Not that there would be any reason to—I was simply an observer and could depart (or nod off) at any time—nor was anyone there likely to care about my opinions, pro or con. But I had that experiential bottle handy, metaphorically clinking around in my laptop bag, nonetheless.

JPL is managed and operated for NASA by an academic institution, Caltech, so it's a somewhat more academic setting than other NASA field centers or head-quarters. These people work hard, and while they can be intense, they are opti-mistic and *collegial*. They are truly on a mission. In other meetings at other NASA centers, while the people there are driven and care deeply about their work, there can sometimes be a sense of tired institutionality. NASA employees can appear to be worn down by the endless procedures and the unrelenting bureaucracy. Not so here. The JPL'ers take to the task with a vigor and sense of purpose that would have inspired General Patton. It's refreshing.

I arrived early, as it can take some time to get through the intense security at the front gate. I signed in and waited while they checked the veracity of my documenta-tion. Webster soon showed up to escort me and we were on our way. Crossing the campus, one always notices that previously mentioned sense of purpose as demon-strated in the faces passing—these people are infused with it. I envy them.

We made our way to the operations center, part of the two buildings that make up the SFOF (reminder: Space Flight Operations Center). Upstairs there are many meeting rooms and offices; we proceeded to the one designated for this morning's meetings.

The room was probably fifty by sixty feet, all in government-mandated gray and white. Not World War II–issue, but definitely decorated with a limited budget in hand. Little gets wasted at NASA, and you can feel it. Even JPL has a tone of spare efficiency—whatever it takes to do the job and little more. The wonders of the universe are yours, but don't expect any extravagance. In the center of the room was a cluster of long office tables, wired up with teleconferencing gear and hookups for laptops. Science in the twenty-first century is conducted via laptops, and to my inner satisfaction the vast majority of these were Apple MacBooks. The perimeter of the room was lined with more working spaces, each adorned with a PC tower (remember those?), most running on Unix. Efficient, if a bit Spartan.

Webster and I settled into a corner, and while I prepared to take notes he nodded hello to a few people. He has been at JPL for over a decade and probably knows more about what is happening on Mars at any given time than any other civilian, yet there is a humility about him that impresses. He's a smart and capable person, about my age but (in stark contrast) tall, lean, and trim. There is little pretension in him—he came to JPL as a newspaperman years ago (when such jobs were far more common than they are today) and stayed on to run public relations for the Mars exploration program. He clearly has vast amounts of respect for the people he works with and lets it be known how much he enjoys the opportunity to sit in on these sessions. Where some in such positions might have been checking e-mail on their phones or gazing out the window, Guy Webster attended to the proceedings with rapt interest, as did I. But he's done it a heck of a lot more times than I have, and I admired his enduring enthusiasm. I've also noticed that a few months later he is likely to remember details from the meeting that have fled my mind. That's the kind of person you want in this kind of job.

A few of the participants for this morning's meeting were already in place, quietly catching up on the dozens—sometimes hundreds—of e-mails that have stacked up overnight. It's a steady stream of mission-related correspondence. They are a busy bunch.

As the rest of them drift in ahead of the start time, I begin to get a bit uncomfortable. Why? Oh, I see—I am in a room full of children. At fifty-seven, I am probably the oldest person in the room (Webster may have a year or two on me—if so, it is much appreciated). The Mars exploration program is run by people who look like grad students with a few premature gray hairs here and there. I know that a couple of them are in their forties, but you'd not guess it if you didn't know—such is the preservative power of having passion for your work.

We were scheduled to attend two meetings today: the first would be a broad daily-report sort of affair, and a couple of the high-level people sit in for that. Ashwin Vasavada, one of two deputy project scientists and a direct report to Grotzinger, was there. The meetings were to be chaired by Bethany Ehlmann, who is in her early thirties and already is an assistant professor in planetary sciences at Caltech as well as a participating scientist on MSL. She radiates unassuming capability.

Daniel Gaines sits to the side. He is the a senior member of JPL's artificial-intelligence group, and today serves as a tactical uplink lead, overseeing what will be sent up to the rover at the end of the current shift, instructing it what to do tomorrow. All of these people have titles, but few constrain themselves to that area. There is just too much to do, too much to learn.

Dan Limonadi sits next to Ehlmann. He's in his early forties, came to the United States from Iran as a child, and quickly realized that his fate rested in space science. He is a lead engineer for surface sampling and a specimen of a man—tall and rugged looking (where do they *get* these people? It looks like an H&M commercial in here). I've spoken to him before, and he is also a pillar of the community outside JPL—he performs all sorts of community good deeds, works on the local search-and-rescue team looking for lost and injured hikers in the local foothills, and coaches some sport or another. I rarely found time to even attend my kid's soccer games, much less coach them. Sheesh. Limonadi is busily preparing for his part of the meeting.

A couple of rover drivers, a deceptively simple term for people who need to understand every detail of how the rover functions while crossing terrain, sit nearby. One is Paolo Bellutta, who will begin the discussion about driving goals for the upcoming sol. Bellutta also distinguished himself by developing software to help in the selection process of a landing site by assessing drivability. The other is Mark Maimone, whom you may have seen on camera if you keep up with JPL's videotaped rover updates. He is one of the few people here who is not rail thin, and I silently thank him for that. But he has a doctorate in robotic sciences and is better looking than me. Oh well.

Beth Dewell is a tactical uplink lead, and her work dovetails with that of Gaines. She looks young and sounds smart, just as they all do.

Youthful though they are, most of these people have experience dating back to the MER rovers Spirit and Opportunity, and some to Pathfinder. They are the senior (in experience, not age) faces of Mars surface exploration post-Viking. And many have doctorates, and if not, technical masters' degrees. That's a lot of education in one small room.

The meeting is called to order—it is not a stuffy affair but rather is collegial and delightfully informal. Nonetheless, it is efficient and the business gets done. The official name for this assemblage is the Science Operations Working Group, or

SOWG. It is not the best acronym at NASA, conjuring up images of something that might end up on my breakfast plate (I'm not a vegetarian). I prefer to think of them as Curiosity's caretakers, minders, and teachers. The machine may be semiautonomous, but a lot of smart people are behind what it does each day—excuse me, each sol. Many of them are here today.

The meeting includes people who are phoning in from remote locations, and it begins with a discussion about Curiosity's current uplink and downlink status, power levels, and other immediate concerns. Nothing jumps out here, and it sounds as if all is well. Of course, I am a tactical-planning-meeting newbie, but Gaines said as much, and everyone seemed pleased, so I take that as a good omen.

Next they talk over the terrain surrounding the rover. Remember that Curiosity can drive autonomously, and do so farther and somewhat faster than its predecessors could. So knowing what is nearby and in its path is pretty important. First it makes everyone's job easier, including the rover's. Second, it avoids any nasty surprises. One of the drivers notes that here is a precipice up ahead. It's impossible from this angle, either from the Mastcam or from orbit, to really glean quite how far it drops, so they are exercising caution. It does not appear to be anything huge, but it does not take much of a drop to cause problems. A hard right-hand turn would provide a safer, more predictable detour, but adds dozens of feet to the drive to Mount Sharp. At these speeds, this kind of distance matters—Curiosity's top speed is 1.5 inches per second, or about 450 feet per hour. That's less than a quarter mile per hour. And at most times it will not be driving near top speed, so this gives you an idea of why distance is important.

But safety matters more than speed. It's at times like this that you realize, despite the challenges and hardships it would entail, just how nice it would be to have a human being up there on Mars. An astronaut driving NASA's lunar rover from the 1970s could drive over far rougher terrain at about 8 mph, topping 11 mph when pushed. It's that human Mark One Cranium computer at work. And, of course, humans have an innate ability to fix things that might go wrong on the surface so far away. You quickly realize just how vulnerable robots on distant worlds are—mishaps can quickly be fatal. Patience and planning are the watchwords.

The group gathers around a pair of monitors. One side shows a 3D model of the terrain surrounding the rover, including a blank patch where the precipice that has them concerned is located—no data. The other shows a photomap of the same terrain.

There is another issue to consider. The area they are crossing is hard ground, like a packed, cemented lakebed. Normally this would be great news because hardpan surfaces allow for faster driving—nary a sand dune or trap in sight. But—and it's a major but—there is a problem with this hard surface. There is a bunch of small rocks all over the area. Normally these rocks, maybe the size of baseballs, would be a nonissue. But they are sharp and potentially dangerous to the rover. Apparently, millions of years of wind-scouring resulted in a bunch of jagged points sticking out of the area like a parking lot nail strip. You know, the ones that are to prevent you from going out the wrong driveway, and if you do, rip your tires to shreds.

Sometime earlier, the techs had been performing the usual "let's look her over" picture sequences with the MAHLI arm-mounted camera pointed back at the rover. Something was not quite right—there were extra holes in the wheels. The wheels are supposed to have *some* holes. There are a series of them, which are used to track wheel revolutions via their imprints in the soil—there are slots and circles machined into them that spell out JPL in Morse code. The lab had wanted to just etch letters, but NASA nixed that, either out of design concerns or possibly institutional envy. So Morse code it was.

But those are cleanly machined holes. Now there were extra dents, holes, and even tears in *all* the wheels. Recall that these wheels are about the diameter of small beer kegs. The rover they support weighs a ton. If they were made thick enough to resist all damage (each was machined out of a solid block of aluminum), they would weigh far too much. So instead, they have the strong areas where the cleats or ridges are, but the connecting metal around the diameter, between those cleats, is pretty thin.

The wheels can take a lot of abuse before failing, but nobody wants to even come close to that point. It's early in the mission, and they have a long way to go. It's just another thing to worry about, another obstacle to quick progress. They will have to drive slower to avoid the worst rocks and minimize the damage. Mount Sharp just got a little farther away.

As this debate is concluded, I notice that one of the guys on my left is typing commands into a computer. So what, you might ask? It got my attention because the information on the screen was just lines of green text on a black background— probably in Unix—and looked just like my first computer, an IBM XT, from 1984. *Déjà vu*, flashback, my goodness I am old. Sigh.

Vasavada caps the debate by saying that the choice of driving route is not a decision for the science folk, it's one for the strategic planners and drivers. The longer route gets the vote, even though going straight would save several sols. Safety first.

On to the next topic. The rover's state of electrical charge is discussed. Curiosity's nuclear power plant is like a trickle charger on a car battery. Good for supplying a continuous few of a little bit of power, but not nearly enough for the real-time operation of a machine of this size and complexity. The nuclear fuel charges the batteries, and the rover runs off them. The state of charge, essentially how much juice is in those batteries, is affected by the amount of work the rover is doing, what instruments are on, how fast and how far they drive, ambient temperatures, heater use, and more. So it's a constant concern. Right now, it's reported that it's a bit lower than anticipated but still in an acceptable range. They agree to keep an eye on it, next topic.

A few more technical points are discussed, and one thing I get out of it is that the current light-time between Mars and Earth (the one-way time that a radioed message takes) is ten minutes, twenty-seven seconds. Interesting.

The first meeting adjourns and Vasavada and a couple others depart.

The next meeting is for more immediate day-to-day operators and the science team representatives. They need to plan more minute details. By golly, I thought that the previous meeting *was* about the minute details. Nope.

Ehlmann conducts a roll call as the people around the table identify themselves for the benefit of those attending online. The meeting kicks off with Ralph Milliken, who is attending remotely and is an assistant professor from Brown University. He looks to be about thirty from his photo; I suspect that he must have received his doctorate when he was nine. . . .

I've become age-obsessed. Moving on.

He refers the team to a topographical map of the area, then discusses the Mastcam test they are trying to complete. Also, there is an outcrop nearby that would be great for the geologists (he's one of them) but probably not so good for the drivers and the rover's wheels. They collectively look at another image of a punctured wheel, and that discussion begins anew.

Then plans for the next few days are discussed. These involve instruments and the tasks needed to operate them at each juncture. Certain tasks require the machine to be at rest, others don't. For tomorrow they plan to do what is called "untargeted remote sensing," which means in effect telling the rover to keep her eyes open and record what she sees as we drive (everyone calls Curiosity "her"; guess I should start as well). Then on the next two working days, Friday and Monday, they plan to do "contact science"—the robotic arm will reach out and touch a rock or soil patch—which obviously requires the rover to be still. During the intervening weekend, there will be more contact science, and the team discusses the high- and low-priority targets.

Once Curiosity has these instructions for a specific sol, conducting the science becomes an autonomous process. Of course, many procedures are monitored from the ground, which is especially important for the people watching the arm when doing contact science. You're placing the arm close to something solid and unyielding, usually a rock of some kind, and you do not want to either bang the arm into the rock or have the rover shift or slip while you are near the rock. You also don't want the arm to hit or scrape any part of the rover while it is being positioned. These sound like minor concerns, but with a one-ton machine sitting behind the arm, any unintended pushback or hard bumping could be disastrous. The onboard software is constantly vigilant and programmed to avoid most potentially dangerous situations, but it cannot know everything. It needs updates for new situations.

The following Tuesday they will use the MAHLI camera to look over the wheels again. Today they noted one new pinhole and small tear on one of the wheels, but nothing showstopping. Just more slightly above routine wear. The wheels can take a lot of abuse, as the surface area in each is so much more than this level of damage can compromise.

A set of rover instrument conditions is surveyed in preparation for another cold Martian night:

Mobility heating—keeping the drive motors happy.
Arm heating—keeping the joints on the robotic arm limber.
DAN passive—the Dynamic Albedo of Neutrons instrument is not working in active mode (i.e., not actively shooting neutrons into the ground), but is passively reading neutron activity in the soil.
Mastcam heating—the cameras on the camera mast will be kept warm.
MAHLI heating—the microscopic imager on the arm turret will be heated.
Navcam heating—ditto.

Ehlmann, who is also chairing this meeting, performs the time-honored tradition of the go/no-go poll of the principals in the room. This process was termed "going around the horn" by Gene Kranz, a flight director in the Apollo lunar landing days. The term was not applied here, but the process was the same. Everyone said "Go."

There is then a detailed discussion, with some dense graphics up on the big screen overhead, of various technical issues. The last data downlink, which gives them, among other things, the information needed to plan the next day's drive, was not as much as expected. Some kind of bandwidth bottleneck. They will plan another downlinking session with the Mars Reconnaissance Orbiter, one of the two orbiting data-relay conduits available.

As mentioned, they will be handing over the rover to the next shift with a bit less power reserve than they would like. They will shorten the drive tomorrow to compensate and move some of the post-drive imaging targets to a later time. Lower than expected temperatures are affecting the batteries, draining them faster than is optimal. But it is all well within acceptable limits.

James Biehl, who joined this meeting after the first one adjourned, talks about the upcoming terrain and driving decisions in more detail. He joined JPL in 2010, and from the looks of him, that was fresh out of grad school. He appears to be a smart and intense kid. He announces that they plan to instruct the rover to drive twenty-six meters tomorrow, or about eighty-five feet. This will be accomplished in a little under an hour at the end of the sol, allowing the rover to perform as much science as possible while it sits in the Martian sunlight. They will snap an image of the wheels frequently to make sure that the drive is not causing any more damage than is absolutely necessary.

A plan for activity with CheMin is discussed, and the restricted data flow comes up again. It should not stop the instrument's activity, but it is on everyone's mind. They look over all the data modules that need to come down from the rover, some technical, some scientific. Priorities are assigned: critical 1, critical 2, etcetera. This way they can get what they really need to move forward and save some of it, mostly science and rover-status data, for a later data-relay pass. Curiosity's computer is, of course, capable of storing information for later uplinking, and the orbiters can do the same if need be.

The meeting begins to feel like a technological ballet. This is the detailed working session, and everything is presented, discussed, analyzed, prioritized, and positioned in an exquisite fashion. It is an amazing synchronization between departments, between instruments on the rover, and between the rover, the orbiters, and the ground stations in the Deep Space Tracking Network. Even the resolution of the images to be downlinked is discussed, as lower resolution leaves more bandwidth for other information. Higher-resolution images can be stored for later if they are later deemed important.

As regards data flow, they discuss windows of availability. The Mars Odyssey pass, later that evening, will allow for seventy-four megabits of information. That's only about ten megabytes, which is not a lot of bandwidth. The Mars Reconnaissance Orbiter will provide more, as it has a larger (almost ten-foot) radio dish and what is called an "adaptive data rate." It can adjust to the needs imposed upon it. Incidentally, noted on the complex spreadsheet-style grid filling the big screen is leftover data from a week ago—there are science people still waiting for results and information at least five days old. They will download it when they can. None of it is critical, but I can visualize some foot-tapping going on wherever some of these scientists who are joining us online are based. Science takes patience, but it's easier to wait when data is being gathered and crunched than when that data are sitting in storage, waiting to be transmitted. But everyone involved is reasonable today.

Another trade-off is chosen—this one limits the Navcam imaging to a spread of 105 degrees instead of wider (this saves on data and possibly electricity.) This limitation makes one of the rover drivers a bit nervous—he's doing his job, protecting his machine. Ehlmann notes with a smile that the decision "probably does not inspire warm and fuzzy feelings." Something has to give here, and the driver accepts the decision. While the Navcams and Hazcams image nearby terrain to determine the safest drive, the Mastcam will spend some time shooting images farther ahead for planning purposes.

ChemCam reports: status okay.

There is a question about the availability for the Mastcam to do a "workspace survey." It is a chance to document the area in which they are currently residing before the drive tomorrow.

It is estimated that the rover will begin tomorrow's operations with a battery charge of "about" 48.27 percent. How they can estimate so exactly is a mystery to me. After all items are considered, they restrict tomorrow's drive to ten meters, or about thirty-two feet, from the previously planned amount, which was almost three times that.

One more go/no-go with these refinements considered. All is still well.

Fig. 21.1. DRIVING TEAM: Curiosity's drivers meet to discuss tactical planning. Second to the left is Brian Cooper, speaking to Vandi Tompkins, in the center. Cooper also designed the software that allows the rover's drives to be planned and implemented. Tompkins creates driving and arm-motion sequences, which the rover then executes. *Image from NASA/JPL-Caltech.*

A meeting later in the day is planned for a smaller group comprised of the science team representatives and a few technical people who will report on tweaks to the planned activities. There is little rest for those involved in this day's planning, and it is no wonder that they rotate people in and out of the most intense "tactical" positions on the mission every few days. It involves an unbelievable amount of work and attention to detail, and it is enervating to simply sit and listen. Actually making the determinations is probably quite energizing, but I bet a few shoulders sag when they break for coffee after the meetings. It is a lot to track.

The second meeting is adjourned, and Webster and I prepare to leave and find the gent who authored the software used to plan and execute the rover's driving schedule. As I watch the assorted participants pack up to return their respective offices or go to another meeting (there is always another happening somewhere), I notice that they are smiling, are chatting, and show little signs of overt stress, at least not between them. Nobody displays any sense of having their toes stepped on, their data constrained, their activities delayed—though many have just experienced it. There is a shared understanding of the challenges, the limitations, and, perhaps most important, *the mission.*

It's really something to see.

CHAPTER 22

PLACE NAMES AND PARANOIA

We are about to go driving with Curiosity across the face or Mars. During this trip, you'll hear plenty of place names. Some will sound pretty normal—Rocknest and Darwin—and others kind of weird—see Glenelg, below—so where exactly do they get these names for places on Mars? It's a question a reasonable person might ask.

In the old days, back in the nineteenth century, there was no official naming system for planetary features. The nationality of the observer and mapmaker often dictated what names were used. An Italian astronomer named Father Angelo Secchi, working at the Vatican Observatory in 1858, gave some of the larger features the names of famous explorers (with a bias toward Italian ones, of course).

In 1867, English astronomer Richard Proctor, utilizing drawings made a few years earlier by a fellow Brit, renamed some of those markings—and additional ones—after astronomers from the past and (then) present. Camille Flammarion, a French astronomer and, oddly, a fanciful mystic, also made maps, this time with geographic names in French. Flammarion was a pretty wild fellow—he drew detailed maps of his observations of Mars even as he wrote early science fiction (rather ahead of its time, but in terms of prose he was certainly no Jules Verne), discussed intelligent life on Mars (no evidence really, just wishful opinion), and penned extensive treatises about the afterlife and universality of extraterrestrial beings. Many thought him to be a bit of a nut.

Giovanni Schiaparelli came along in the mid-1800s and made his best observations and maps in 1877 during a close approach of Mars. He, of course, used Latin names for his maps—he was Italian, after all. He was also one of the first to create maps of the surface of that planet in detail. Much of the detail he charted was illusory, probably a result of eyestrain and a vivid observational imagination, but those interested in Mars were sufficiently hungry for *any* detail that his maps and nomenclature had substantial staying power.

Other astronomers took turns at the eyepiece over the next few decades and a few tried their hand at renaming things, but when Percival Lowell came along in the 1890s and built his Mars-only observatory out in the Arizona territories—devoting his copious spare time to observing that planet (he was independently wealthy, as you will recall)— he adopted Schiaparelli's naming tradition and it stuck. Mars would be labeled in Latin, and these maps were used in one form or another until the mid-twentieth century.

Thank goodness for the International Astronomical Union (IAU), a global federation of academicians, who stepped up and in 1960 standardized a naming scheme for Mars and other bodies. Here is its schema (in my terminology):

LARGE CRATERS: Dead scientists and writers who contributed to the science and lore of Mars;

SMALL CRATERS: Global villages with a population of less than 100,000 persons;

LARGE VALLES (Valleys): The names for Mars or the word "star" in various languages (but not always—Valles Marineris was named after the Mariner spacecraft, for example);

SMALL VALLES: Classical or modern names of rivers;

and so on.

You get the idea—it was a weird but generally standardized approach. By the time that NASA's robots got to Mars, the tradition was well established and the new features being observed were worked into the existing maps, which had been made from Earth-based observations. But by 1971, the images were getting clearer, and more craters of all sizes, along with valleys, volcanoes, and myriad wind- and water-sculpted features were photographed. Still, classical names were applied.

This held up for the most part through the Viking era, with some informal terms used for terrain near the Viking landers. Apparently the IAU decided that items smaller than about 330 feet should not receive official recognition (i.e., names), so the door to the barn was left open and the metaphorical cows scattered.

The Viking landing site included features with such names as:

Big Joe
Bonneville

Delta
Midas Muffler
Mr. Badger
Mr. Moley
Mr. Rat

There is a theme here: playfulness, whimsy, and classic literature. The last part is clearly a generational theme, as we will see next.

Now, look out! It is 1997 and here comes Mars Pathfinder, with its crew of young and apparently irreverent scientists. Prepare to be amused/horrified (depending on your inclination).

As objects near the Pathfinder landing site were imaged, they got a whole new set of names . . . ones that would not tickle the IAU people, if they were in the room, any more than "Midas Muffler" likely did. The region in which Pathfinder set down was called Ares Vallis ("The Valley of Ares," after the Greek name for Mars) and that moniker was observed. But not what was *in* it. A partial list:

Barnacle Bill
Yogi (after Hanna-Barbera's Yogi Bear cartoon character)
Scooby-Doo (also a Hanna-Barbera creation)
Anthill
Baby Otter
Chimp
Goldilocks
Gumby (after the wacky 1960s kid's TV show)
Jimmy Cricket (from the early Walt Disney movie *Pinocchio*; the actual char-
    acter name was "Jiminy," but apparently this was either lost on the excited
    young engineers, or "Jimmy" was intended as a nickname)
Lumpy
Nibbles
Snoopy
T. Rex
Zorak (another Hanna-Barbera 1960s character—a man-sized praying mantis
    with a lisp from the largely forgettable cartoon *Space Ghost*)

And others. See an age-dependent trend here? The Pathfinder team was enjoying the mission like crazy, and it showed. Some in the NASA hierarchy were less amused, but these were unofficial names, so what the heck. They were having a blast.

Then—someone at a gray metal desk woke up. This person may or may not have been a part of the legal office of the space agency, but for reasons that are difficult to pin down with authority, a new rule came about: no names were to be used that might be subject to copyright law. That proscribed just about anything created within the last seventy years, so at the time of the MER rovers that flew in 2004, this meant no creations subsequent to 1934. And as we know, some properties created earlier (think Mickey Mouse) will have copyrights extended come hell or high water, and legally defended with *vigor*. It's not entirely unfair, either, as NASA protects the name, likeness, and context of the use of its own logo—you can't use it to promote a product or make money. So these things cut both ways. You don't ever want to get on the wrong side of Disney and other big-name copyright owners.

Gone were the fun, cool names from our respective childhoods.

In were classical references and homages to important people. Oh, and the geologist's favorite rock formations *yawn*.

Fig. 22.1. WHERE'S BAMM-BAMM: How many potential copyright violations can you find in these rock names? This is a small portion of Sojourner's backyard on Mars. Midway into the MER mission, they stopped using names that might cause legal concerns . . . and Mars became just a bit less fun. *Image from NASA/ JPL-Caltech.*

So why the heck did I take you on this diversionary jaunt? Because a lot of people wonder about the names used to label the features found by Curiosity in Gale Crater. The first place Curiosity explored was called Yellowknife Bay. It was named after a region and capital city in the Northwest Territories of Canada. The earthly Yellowknife is a tiny place, with a population of about twenty thousand, and not exactly Las Vegas. But it has its attractions. According to John Grotzinger: "What is the port of call you leave from to go on the great missions of geological mapping to the oldest rocks in North America? It's Yellowknife." The town in Canada, that is.

So this is apparently what happens when you institutionally crush whimsy and put the geologists in charge. But at least it's a naming system.

For Curiosity's actual landing site, they made some exceptions. The much-beloved science-fiction author, Ray Bradbury, died just a few weeks before Curiosity landed on Mars, so in a gesture that warmed many hearts, including mine, they named it Bradbury Landing. Of course, they could have named it Percival's Pinnacle or Giovanni Gorge, I would have been okay with those, too, but nobody asked me. I expect that oversight to remain uncorrected.

Soon another name bubbled into the awareness of those casually following Curiosity's progress: Glenelg. Huh? How do you pronounce that? Newsrooms far and wide had to stop, sound it out, and run to *Wikipedia* to find out what the heck that name referenced. Here is what they found . . . enjoy:

"Glenelg (Scottish Gaelic: Glinn Elg, also Gleann Eilg) is a village and civil parish in the Lockalsh area of Highland in western Scotland. The parish covers a large area including Knoysart, North Morar and the ferry port of Mallaig. At the 2001 census it had a population of 1,507. The smaller 'settlement zone' around the village had a population of 283. In 2011 Highland Council estimated that the community of Glenelg and Arnisdale had a population of 291." (This comes from the "Glenelg, Highland" *Wikipedia* entry, accessed March 12, 2013. You can find it here, if you are really that curious: http://en.wikipedia.org/wiki/Glenelg,_Highland.)

Huh. That's not much better than where we started. Some science bloggers caught the fact that the word is also a palindrome—it's pronounced the same way when read backward as it is when read normally—and that should have been hot fodder for the conspiracy theorists, though it did not seem to protrude into that sphere, thank goodness. I can hear it now . . .

"*Glenelg* is *glenelG* spelled backward!!! And Curiosity is a rover, and Rover is a dog's name . . . *dog* spelled backward is *God* . . . and so—" you get the idea. These lonely people never tire of finding mystery in clarity. It's the reverse of science, taking logical observation back into the Dark Ages: the "Face on Mars," pyramids on Mars, and underground cities on Mars *and* on the moon. NASA is supposedly run by unrepentant Nazis, Freemasons, and, of all things, worshipers of ancient Egypt. I kid you not, look it up. Oh, and by the way, we never went to the moon. . . .

In truth, Glenelg was named after both a rock formation near Yellowknife in Canada and the previously referenced Scottish region. But according to a NASA statement: "The science team thought the name Glenelg was appropriate because, if Curiosity traveled there, it would visit the area twice—both coming and going—and the word Glenelg is a palindrome. After Glenelg, the rover will aim to drive to the base of Mount Sharp." So, Glenelg it was. They could have also named it "Pop," which would have accomplished the same thing as a palindrome, but once again, nobody asked me. I'm available by request.

A note on the conspiracy lovers: while they may not have made much hash over Glenelg (at least I didn't find it if they did), they have found plenty of other things to get excited about. So far, Curiosity has found—in some people's fevered imaginations—a jackknife, a flower, a person—possibly a Greek figure, Bigfoot, or Elvis (depending on whom you listen to)—miniature pyramids, and many "ventifacts"—objects that are just *too* regular, *too* right-angled, or *too* geometric to be found in nature. We are not talking crop circles here (though if you are a True Believer, I can't see why they would not be on Mars as well)—we're talking rocks (of which there are billions upon billions on Mars—the place is a desert after all) that have "unnatural" shapes. You can look for yourself—the websites usually have titles that include "THE TRUTH ABOUT [insert your favorite space program here]!" in all caps. Or, try a search for "NASA conspiracy" or "Mars NASA truth" . . . you'll see the rocks in question. At best, to my eye, they look like broken ashtrays (wait—that's not natural!) or smashed plaster chunks (ditto).

If this intrigues, do yourself a favor and visit Death Valley sometime. I'd take you there, but I already went and am just happy to be alive. But spend an afternoon squatting in just about any area where there are rocks that have tumbled down from higher slopes. Guess what? You will see rocks that look like flowers, lizards (though it might actually be a lizard—check before touching), swastikas, Paul McCartney, and Osiris. I swear—it's like looking at clouds for too long, and they can be very convincing until you remind yourself—they are just, simply, rocks. Then go back to the aforementioned websites and stare at the photos again, and I do believe that you will see . . . just rocks. Give it a shot if you are so inclined.

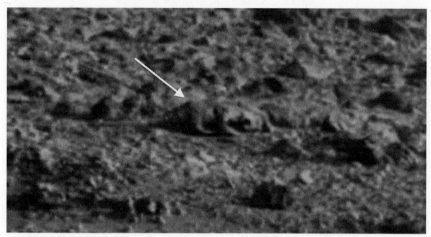

Fig. 22.2. THE MARS RAT: Yes, you could be excused if you took a second look . . . it *does* look like a rat. But enlarged and enhanced, it's just a rock. *Image from NASA/ JPL-Caltech/MSSS.*

If you still see a ziggurat or zeppelin on Mars when you've completed this experiment, let me know. I know a few thousand scientists who would love for it to be true.

CHAPTER 23

ON TO GLENELG

After the August 5 landing, Curiosity sat at the recently christened Bradbury Landing as the folks on the ground got their bearings, checked out the rover and the surrounding terrain with the cameras, and prepared to wake up the rest of the instrumentation aboard.

From orbit they were also able to survey the neighborhood a bit. The heat shield came to rest almost five thousand feet away. Almost opposite, about two thousand feet away, were the backshell and parachute. To the north, about 2,100 feet from the rover, was the final resting place of the descent stage. And of course, they pinpointed the location of Curiosity—so close to the target that the offset was almost equivalent to a rounding error.

Within ten sols, the checkout of the rover and its instrumentation, as well as the upload of the ground-operations software, were complete. Concurrently, the science teams had been busy as well, deciding where to go first. Mount Sharp beckoned from a few miles away, but spectacular as it was, it was not the first target on the minds of the geologists. They had decided, based on new data, to go in the opposite direction.

A closer inspection of what was now called Glenelg revealed some tantalizing geology about a quarter-mile distant. The area showed an overlap of three kinds of terrain and was just too juicy to resist, even though it meant heading away from Mount Sharp.

"With such a great landing spot in Gale Crater, we literally had every degree of the compass to choose from for our first drive," said John Grotzinger. "We had a bunch of strong contenders. It is the kind of dilemma planetary scientists dream of, but you can only go one place for the first drilling for a rock sample on Mars. That first drilling will be a huge moment in the history of Mars exploration." An understatement for certain.

Fig. 23.1. A FACE ONLY A MOTHER COULD LOVE: Glenelg, as seen from the Navcam, is a place only its mother—or a geologist—could love. But it represented a union of three geological regimes in one spot and was therefore too rich to pass up even though it meant heading away from Mount Sharp, Curiosity's primary destination. *Image from NASA/JPL-Caltech/MSSS.*

But why Glenelg? According to Grotzinger, it represented "the union of three interesting geological sites." The entire region was touched by an alluvial fan, and as such would give a variety of rocks and soils from regions above, with water alteration almost ensured. But within this, Glenelg appeared to be special. Eventually called "the Promised Land" by the science team, Glenelg contained what appeared to be layered bedrock, which would be a great place to begin investigations. There was a light-toned area with well-developed layering, which would likely turn out to be sediments, and therefore close to the heart of sedimentologists like Grotzinger. There were also darker, almost-black bands of something running through the area, which were only visible to the scientists once they had eyes on the ground. Adjoining this was a region with more cratering, which could represent older material, and the third type of terrain appeared to be much like the landing site. Once at Glenelg, they would have access to all three in one spot.

Grotzinger again: "We really do think it's going to be interesting. You can see a light-toned unit [of rock], which we believe to be material that from orbit has this signature, this property, of having relatively high thermal inertia. That's the ability of a material to retain its heat. So, late in the day or at night, orbiters are able to observe this rock well because it seems to be giving off heat late into the day and night. We don't know what the reason for that is, but it's always been a bit of a beacon for us and we're getting closer and closer to it." High thermal inertia could mean cemented sediments, and soon Curiosity would examine the region for ground truth.

While preparing for the first drive, they wanted to check out one of their most valuable noncontact instruments. ChemCam would not have to touch rocks to understand what they were composed of, and it promised to revolutionize Mars exploration—as did so much of what Curiosity carried. Rather than drive up, inspect, and "sniff" every interesting target they saw, or merely wish that they could get closer to something too high up to reach, ChemCam would allow the science team to aim the laser at rocks up to twenty-five feet away, which is real time-saver in an environment with interesting targets all over the place. As you will recall, the built-in spectrometers would then read the laser-induced incandescence off the rock to determine its composition.

On August 19 came the first chance try it out. Targeting a 2.5-inch rock they called Coronation, ChemCam was powered up for its inaugural use. Over a ten-second time span, the powerful mast-mounted laser shot the rock thirty times with a brief but powerful one-million-watt blast of light. This generated enough heat to flash burn a tiny layer of the rock that the spectrometer was able to read successfully.

Coronation was nothing particularly special as Mars rocks go—the geologists had thought it was probably a little chunk of basalt, or lava-based rock. But it was handy and sported a sufficiently clean surface to use as a good first target. The resulting spectra showed that it was in fact a basalt and that the instrument was working perfectly.

Curiosity prepared to move out, but not before Coronation gained a bit of additional notoriety. As with Curiosity itself, Coronation was given its own short-lived Twitter feed. The tweets included some typically geeky humor, this one at the expense of Mars's largest volcano, Olympus Mons: "Did you know I was born in a volcano? Basalts like me come from lava. That's why we call it *Olympus Mom*." Ouch. But all's fair in educational outreach, I suppose.

After initial tests—driving, turning, and backing up—the engineers and scientists carefully examined the imprints of Curiosity's wheels in the Martian soil. Besides giving the soil-dynamics people an idea of the consistency of the dirt and hardness of the surface, the tracks were also an indication that all six wheels were working properly. After a few more baby steps, it was time to depart.

On August 22, wheels rolled and Curiosity departed Bradbury Landing . . . although "departed" might be a bit of an overstatement. That makes it sound like the luggage was strapped on the trunk, the bride and groom jumped in, and off Curiosity drove, cans clattering off the back. It was not quite like that. At the slow speeds the rover drivers started with, it was more like defining an embarkation upon a world journey by getting out of your recliner. But this was the commissioning drive, and with a two-year mission at minimum (this is the "primary mission," everyone hopes for much longer), there was no reason to rush. Including the mobility tests, they moved about twenty feet the first day.

The machine moved in apparent fits and starts for almost a week, testing and evaluating both itself and the surface it was driving on. With the long back-and-forth of information coming down, and commands returning to Mars, progress was slow at first. Nobody wanted to rush things—and for certain nobody wanted to make some kind of mistake *now*. They had survived the treacherous landing; an error now would simply be too embarrassing.

During this time Curiosity also had a couple of major media moments. First, a laudatory statement by the NASA administrator Charles Bolden was transmitted to the rover and played back to Earth. Then, in a savvy PR move, JPL arranged for the pop singer and STEM (science, technology, engineering, and mathematics) education advocate will.i.am to do the same thing with a song he composed titled "Reach for the Stars." Kids loved it.

In the first week of September the rover was able to make solid progress toward the first destination. During breaks in driving, controllers tested both the

CheMin and SAM instruments. These tests were conducted with both instruments empty, as no Martian soil had yet been sampled. In fact, the SAM test was performed on residual Earth air, more of which remained in the sample container than had been expected. But everything worked, and that was the important thing. Reading Earth air also gave the science team verifiable readings they could compare to ground tests.

While reaching Glenelg was the primary goal, no self-respecting geologist would give up an opportunity to stop and look at an object of interest. Said object popped into view on August 20, and the location looked like a good spot to calibrate some of the instrumentation as well.

"The science team has had an interest for some time now as we're driving across the plains to find a rock that looks like it's relatively uniform in composition to do some experiments between ChemCam, which we've been acquiring a lot of data with, and APXS, which we haven't used yet on a rock," said Grotzinger. "Both of those instruments could make a measurement and there could be differences between the measurements because one is measuring at a small scale (ChemCam) and one is measuring at a larger scale (APXS). These rocks that we drive by on the plains here that look dark, they probably have basaltic composition. That's a familiar material to us," and therefore a perfect baseline to calibrate the two instruments to one another.

The rock was about the size of a tissue box and roughly pyramidal in shape (don't get excited, conspiracy theorists . . . the resemblance is slight at best and it does not have the Eye of Providence—as seen on the US one-dollar bill—on the side). The science team named it Jake Matijevic in honor of a JPL engineer who died shortly after Curiosity landed. The attribution touched everyone involved with the program.

It took four days to complete the activities at Jake. To get a calibration between ChemCam and the APXS instrument readings, they used both those devices on the rock.

Grotzinger commented on the activity: "The hope is [that] we can analyze this rock and then do a cross comparison between the two instruments," he said. "Not to mention it's just a cool-looking rock there, sitting out on the plains with almost pyramidal geometry, so that's kind of fun as well." A little whimsy in science never hurts.

When the evaluations were done, both ChemCam and the APXS found surprising, and matching, data from the rock. It was unique among the Martian rocks seen to date: an igneous rock, high in the mineral feldspar and lower than expected in magnesium, iron, and nickel. The feldspar indicated that it had been formed in the presence of water, but this was a volcanic rock, and it was closer in composition to terrestrial igneous rocks than anything previously seen. This composition pointed to some ancient and large-scale process involving subsurface, high-pressure volcanic activity and water. These types of rocks are rare even on Earth, and when found here on this planet, they tend to be in oceanic island environments like Hawaii. It's an interesting mental exercise to extend those possibilities to Mars.

Over the next few sols they drove and drove, slow but steady. Then on sol 39, an object came into view that was a showstopper. It was named Hottah and was just under four hundred feet from Bradbury Landing.

Now, if you stumbled over this rocky formation on the way across, oh, say Death Valley (no, I am not obsessing about that awful place . . . I'm not, I'm not), as a civilian, you'd probably not think of it as anything other than a toe-stubber. A geologist would see it as an interesting indicator of water flow in the past. But on Mars? For the geologists, it was like coming across a rusted Chevy, albeit a bit more expected.

To be fair, they had seen something similar a couple of sols earlier that they named Link, but it was not nearly as exciting—or cool looking—as Hottah. And of course they were both named after geological sites of historical note in Canada. You can assume that naming convention from here on unless otherwise noted.

When I saw Grotzinger a day or two later at a media event, he looked (a) like he hadn't slept for some time (they were still in the thick of Mars Time) and (b) like the challenges to his circadian rhythms did not matter one bit. While as calm as usual on the outside, there was a sense of adrenaline coursing underneath the surface. "Hottah looks like someone jackhammered up a slab of city sidewalk, but it's really a tilted block of an ancient streambed," he said in a September 27 press conference. "In some cases, when you do geology, a picture's worth a thousand words." He was pretty damn happy.

That was an apt statement for something as cool looking as Hottah. While I have said it would look downright plain on Earth, even the geological laymen among us could see that this was spectacular. It was a broken and tilted slab made up of water-transported rocks, pebbles, and sand, ranging from dust-sized to as large as a golf ball, sticking up at an angle. These pebbles had been rounded off by long-distance movement—they looked like tiny river rocks, which they were. I'd seen a lot of sedimentary layers in various geology field trips in college, and this was an immediately familiar sight. The bits were far too large to be moved here and eroded by wind, this was deposition by water. It was a nice moment.

The science team's assessment was that this conglomerate layer—a bunch of sand and pebbles rolled along until smooth—had been formed by a rapidly flowing stream, perhaps knee- or waist-deep. While water had been inferred in many locations on Mars and sampled as ice near the poles, this was the first close-up observation of this kind—of the direct result of erosion and deposition by a flowing stream.

Gorgeous as it was, however, it was not the Grail: "A long-flowing stream can be a habitable environment," Grotzinger said. "But it is not our top choice as an environment for preservation of organics. We're still going to Mount Sharp, but this is insurance that we have already found our first potentially habitable environment."

Individual members of the science team would continue to pore over the images of Hottah, taken from a distance by Mastcam's telephoto lenses, but it was time to continue the trek. Suffice it to say that the results were interesting and that flowing water was a comparative slam dunk. There's better to come.

The next stop on Curiosity's tour of a once-wet Mars was named Rocknest. The area was made up of loose soil and appeared to be a prime target for the first sampling activity of the scoop on the robotic arm. The rover proceeded to a part of Rocknest they called Ripple, which was exactly that, a rippled bit of sandy soil apparently deposited by wind. It would provide nicely sifted fine grains.

But first things first. To get to the fresh stuff, Curiosity scuffed some of the dirt by spinning a wheel in it. Then the MAHLI instrument and the APXS were used to ascertain what exactly it would be grabbing. On October 7, sol 61, the scoop grabbed a small soil sample. This and a succeeding sample were used to cleanse the sampling mechanism—the scoop and collection chamber. By taking the sandy soil and moving it around—by moving the end of the arm and vibrating the chamber—it should scrape any earthly residues out of the sample-collection system. This was critical to getting "clean" readings from the incredibly sensitive instruments in the belly of Curiosity.

While this was being performed, someone noticed something in an image of the ground nearby the rover. There was a tiny bright spot. Bright areas had been seen before by the MER rovers, but nothing quite this well defined. What could reflect light like that? Mica does, and it's one of those water-indicating minerals, but nobody really thought it would be that. In any case, it was too peculiar to pass up. The presampling activity with the arm was halted while the science team looked closer.

Fig. 23.2. RIPPLE: At Rocknest the rover gathered its first soil samples at a site called Ripple. Before the rover dug into the soil, it was commanded to drive across some soft areas to assess the soil dynamics—much can be learned just from looking at the wheel tracks. *Image from NASA/JPL-Caltech/MSSS.*

After some head scratching, it became pretty clear that the half-inch-long item was probably a piece of plastic that had flaked or fallen off the rover (it looked like a toenail clipping to me, which would have been way cooler than the truth). That was a disappointment, but one gets used to these things. There had been a similar experience early in the Opportunity rover's life, when it spotted a small object that looked for all the world like a rabbit head, about two inches long. And by that I mean that it literally looked like the logo of the Playboy Bunny. The next time they imaged it, the thing had *moved*. Now *that* was interesting. A spectral analysis confirmed that it was a chunk of soft material from the rover that was light enough to be blown around. Not as sexy an answer as one might hope for, but the likeliest one. Same with Curiosity's bright item: just another bit of manmade Mars litter.

With the toenail mystery behind them, the "arm drivers" gathered a third and a fourth sample of Mars dirt. These samples were placed, one after the other, into CheMin. The rest of the fourth sample was used for more shake-and-scrub of the sample-collection device, as SAM was so sensitive that it was feared there might still be contamination to remove.

Also of note, there is a little tray on the front of the rover where the sampling mechanism can drop a bit of the material before it is sent onward to the instruments within. When soil is placed on this tray, it provides a clean metal background against which the MAHLI instrument can take a nice, controlled close look at some of the sample, allowing the scientists to decide what the next step should be.

The sample for CheMin needed to be about the size of only a baby aspirin. CheMin was able to receive more than one of these samples at a time due to the clever nature of its design—it has seventy-four separate containers that rotate into position on a wheel, fifty-three of which are able to accept samples. Each of these has clear, flat, glass sides that allow the high-energy x-rays to pass through and do their work. The resulting image, as described earlier, is a diffraction pattern that shows visible signatures of various minerals.

Within a sol, the results were in from CheMin: the soil sampled at Rocknest was similar to weathered basaltic soils (of volcanic origin) found in places like Hawaii. The remainder of the sample was made up by materials like simple glass. You could not pick a less Hawaii-like place to find this soil, except perhaps Jupiter or the sun, but there you have it.

As David Bish, the coinvestigator on CheMin said in a NASA news release, "Much of Mars is covered with dust, and we had an incomplete understanding of its mineralogy." He added, "We now know it is mineralogically similar to basaltic material . . . which was not unexpected. Roughly half the soil is noncrystalline material, such as volcanic glass or products from weathering of the glass."

So what about the ongoing quest for water? "So far, the materials Curiosity has analyzed are consistent with our initial ideas of the deposits in Gale Crater," Bish continued, "recording a transition through time from a wet to [a] dry environment. The ancient rocks, such as the conglomerates, suggest flowing water, while the minerals in the younger soil are consistent with limited interaction with water." This sounds like a lot of water way back when and a lot less in the last billion or so years . . . which matches hypotheses generated by other investigations by the MER rovers and the orbiters.

Soon the SAM instrument stepped up to the plate. While it had not yet ingested any soil samples, it did grab a sample of ambient atmosphere. The SAM team members soon realized that they had, at last, some real data on what the heck might have happened to Mars's atmosphere so long ago that helped to turn the planet from a wet Eden to the Sahara. The concentration of heavier isotopes of elements of carbon, as compared to that found in ancient rocks, demonstrated that the lighter versions of carbon isotopes had, over time, drifted away from the planet. This and similar results later in the mission will hopefully be validated once NASA's newest Mars orbiter, MAVEN (Mars Evolution and Volatile Atmosphere), arrives late in 2014. MAVEN will try to track the present-day loss of lighter elements from the Martian atmosphere in an effort to project backward and understand what happened in prior epochs.

Finally, in early November, the SAM instrument was able to taste Martian soil. Some of the leftover CheMin sample (remember that they are very small) was fed to the waiting SAM instrument.

Another benefit of the rover's design was its ability to hang onto recently grabbed samples even as it moved off to other sites. Not only could some soil be kept in the scoop for a while but the sample chambers in the instruments could be rotated back into the active area of the machine more than once. It would take weeks, but more data would come from the Rocknest samples.

In the interim, the rover moved toward an overlook called Point Lake to use the Mastcam to map out routes for the next sequence of drives. On the way, it used the APXS for a couple of "touch-and-go" quickie examinations and fired the ChemCam laser for its analysis.

Also during this time, a dust storm kicked up on the planet and was first spotted from orbit. Curiosity noted a consequential shift in atmospheric pressure, and between it and the orbiters, scientists were able to track the storm's progress. It lasted over two weeks. While this Martian gale never got very close to those experienced by either Opportunity or Curiosity, another type of storm was brewing. It too involved Mars, but this one took place on Earth and centered on JPL itself . . .

CHAPTER 24

"ONE FOR THE HISTORY BOOKS . . ."

What a difference a few words make. Specifically, the five in the title of this chapter. The small firestorm these few words caused—a media sensation in terms of planetary exploration—were way out of proportion to the content. The media went wild for a few days and then embarked on a predatory, extended three-week watch over JPL and NASA.

Up front: John Grotzinger is a smart man. By the end of any of our hour-long sessions together when I was interviewing him for this book, my brain hurt. It's that much work just keeping up, and he was taking it easy on me. As I have meekly mentioned here and there, I have some basic education in geology, which served me well for all of about ten minutes . . . then we would get to the material that felt more appropriate for the graduate students. Or at Caltech, the undergrads . . . in the first session of the first class. You get the idea. Thankfully, as a writer all I need to do is take notes and try to ask seemingly intelligent questions, based on what I had learned in the preceding discussion. But even that could leave one gasping.

That's one writer's point of view. There are a lot of very smart science writers out there, and one of them is Joe Palca of National Public Radio (NPR). His coverage of the MSL mission was as good as everything else he does—which is really good. But quality does not necessarily sidestep controversy, often quite the opposite is true, even if that controversy is unintended. And here is where the tale gets interesting.

In late November 2012, about four months after Curiosity began surface operations, Palca filed a story on NPR. In it, he narrated and produced an edit of the following exchange. He had earlier spent an afternoon with Grotzinger in his JPL office conducting an interview on mission progress. The SAM instrument had ingested the first sample and had done its job.

On NPR's *Morning Edition* of November 20, 2012:

Palca: "The lead scientist for the Mars rover mission is John Grotzinger. I interviewed him in his office at the Jet Propulsion Laboratory in Pasadena last week. While I was setting up my equipment, Grotzinger was glued to his computer screen."

Grotzinger's voice: "We're getting data from SAM as we sit here and speak."

Palca injects a verbal definition of what SAM does.

Grotzinger continues: "The data looks really interesting. The science team is busily chewing away on it."

Palca then says to the audience, "Right now, SAM is working on a Mars soil sample, and Grotzinger says the results are earth-shaking."

Back to Grotzinger's interview: "This data is going to be one for the history books. It's looking really good."

The interview continued, but that's the important part.

To be fair to Palca, the exchange was apparently reported accurately. It was a good story on the events of the day. What got everyone's attention were the statement "Grotzinger says the results are earth-shaking." Then there were Grotzinger's words, "This data is going to be one for the history books." It was not said in a way that implied anything more than good, exciting science.

But that's not what the rest of the media heard.

In a feeding frenzy that built on itself for the next twenty-four hours, the headlines grew from this simple story like a Chia Pet on steroids, or if you were a part of the NASA PR machine, perhaps more like an outbreak of the Ebola virus. *Popular Science* online said: "What 'Earth-Shaking' Evidence Did the Mars Rover Curiosity Find?" Many other outlets took the statements badly out of context; some outright embellished them. A few headlines misquoted even Palca, saying "earth-shattering," which is somewhat different.

More placid accounts, such as those from Space.com, used language more befitting the situation, like "Mars Mystery," or "Has Curiosity Rover Made Big Discovery?" which at least leaves the main points open to the reader's speculation without breathlessly churning the facts or deliberately misquoting anyone.

Understandably, to anyone following the mission with a modicum of attention, when someone talks about a major discovery (not Grotzinger's words, mind you) with the SAM instrument after a soil analysis, the first thing that sprang to the minds of many was a possible indication of organic carbon. Certain isotopic numbers can indicate organic forms of carbon, and most of us would love for that to mean a bio-logical origin. But that's a long leap, and not what was being said here at all.

Soon other outlets were trumpeting headlines like "So, What Is NASA Curios-ity's History-Changing Discovery?" (from the Wire via Yahoo News) or News.com's "Life on Mars? Curiosity Rover Stirs Excitement." The *LA Times* said it thus: "NASA Dangles Big, Secret Mars Discovery, but We Don't Want to Wait." Really? I certainly didn't perceive any "dangling" occurring.

The public-relations operation at JPL now had a problem. Publicity is great, and planetary exploration can use all the help and recognition it can get. But nobody wants to be misquoted, or even mis-contextualized. Though it was nearly sixteen years before, a lot of people remembered the now-infamous news confer-ence about the Martian meteorite from Antarctica that had *appeared* to have possible fossils of microorganisms preserved inside. While the case is still open in some peo-ple's minds, the announcement was sufficiently premature by scientific measures that it did not take long for dissenting voices to say "Not so fast!" and eventually the story died an ugly death (the recent discovery of some possibly biologically caused water migration in another Mars meteorite has reopened the question). Nobody at JPL, Caltech, or NASA wanted a repeat of that fiasco.

Within days, Guy Webster was quoted as saying, "It [the story] won't be Earth-shaking, but it will be interesting . . . the whole mission is one for the history books." It was an accurate assessment, and a valid attempt at informing the public without getting bogged down in details that needed to be delivered at length by the science team. But that is not the message most people wanted to hear (espe-cially the clamoring journalists), and in the inevitable backlash, headlines such as this from the *International Business Times* came along: "Mars Rover Curiosity's 'Huge Discovery' a Dud . . ." It cited NASA as "breathlessly announcing" a discovery by Curiosity and cited Grotzinger as the source of "hype," which was patently wrong and unfair, across the board.

Unfortunately, NASA went quiet on the story at that point. Informal discus-sions with concerned parties at JPL indicated that they had been told, as a group, to not discuss the story for the time being.

Grotzinger, understandably, stayed away from the microphones for a while. An announcement was made that the "discovery" would be discussed at the American Geophysical Union's conference in San Francisco on December 3, 2012. The date was not very far away, but the wait was agonizing, even though expectations had been dampened. The press, with its ever-voracious twenty-four-hour news cycle, does not like being told to *wait*.

Full disclosure: within days of the first announcement, I inadvertently waded into this steamy territory. I had already shot a long video interview with Grotzinger, and before releasing it added a comment about a possible mystery within the Curiosity mission's findings—it was within that first twenty-four hours and my brain also whispered "Could it be organic molecules?" While my media statement was not technically inaccurate as such, the timing could have been better—and when I realized the immensity of the "history books" brushfire, I was not thrilled at being even the smallest part of it. I was also writing about Mars for the *Huffington Post* in the same time frame, and I filed a story about this subject somewhat later, but my angle was more focused on the AGU conference and why NASA had treated the story the way it did leading up to the conference. End of disclaimer.

On November 29, the outside world got a tidbit from JPL's media operation:

> PASADENA, Calif. —The next news conference about the NASA Mars rover Curiosity will be held at 9 a.m. PST Monday, Dec. 3, in San Francisco at the Fall Meeting of the American Geophysical Union (AGU).
>
> Rumors and speculation that there are major new findings from the mission at this early stage are incorrect. The news conference will be an update about first use of the rover's full array of analytical instruments to investigate a drift of sandy soil. One class of substances Curiosity is checking for is organic compounds—carbon-containing chemicals that can be ingredients for life. At this point in the mission, the instruments on the rover have not detected any definitive evidence of Martian organics.

This was the most we'd heard since the whole dustup began, and it was a pretty conclusive announcement: there will be an MSL panel presenting results of the SAM analyses, and it's not anything "earth-shaking" (remember, those were Palca's words, not Grotzinger's). Fair enough—take a deep breath and regain some composure, media people.

On December 3, the conference began, and a gaggle of press converged on the Moscone Center in San Francisco. Wisely, the conference's organizers had placed the Curiosity panel's main presentation at the front of the program. Once the panel had concluded an hour later, the press thinned dramatically—think of it as media attention-deficit syndrome. Everyone had seen the previous week's announcement from JPL but wanted to hear it from Grotzinger himself, and they did.

Michael Meyer from NASA headquarters began the session by giving a capsule description of the mission's progress since landing. Our friend Ken Edgett then described the sample-gathering activities in some detail. Ralf Gellert, who works on the APXS instrument, discussed the sand drift (at Rocknest) that had been sampled. Then the moment so many had been waiting for was nigh: Paul Mahaffy, the principal investigator for SAM, spoke about the test results.

"We really consider this a milestone, the two instruments buried inside Curiosity got their first gulp of Mars material," he began. "It really is a rich data set." He talked about the significance of both SAM and CheMin, as the reporters leaned forward in their seats. What was the announcement? "I'm going to talk about the highlights of the SAM data, the volatiles or gases released from these samples." Ah, here it was. Pens were poised above writing pads in anticipation, video cameras were rolling.

Fig. 24.1. STRAIGHT TALK: After two weeks of out-of-control press reactions, John Grotzinger sets the record straight about what Curiosity "found" on Mars. The early results of the SAM analysis of the soil did indicate organics, but it could not be confirmed that they were of Martian origin—they might have been earthly contaminants or organic material from a stray meteor. The analysis would continue, and he counseled "patience." *Image from JPL-Caltech/NASA/AGU.*

"Let me start off by saying: SAM has no definitive detection to report of organic compounds with this first set of experiments." Drat.

"It's not unexpected that this sand pile would not be rich in organics, as it's been exposed to the harsh Martian environment, and we really selected the site in order to scrub out the sample-processing system. . . ." Okay, so the soil had been baking in Martian UV for eons, and was selected as much for its cleansing properties as anything else. Got it.

He talked a bit more about how both atmospheric and soil samples were selected, ingested, and evaluated. He then discussed the relevance of isotopic numbers in test results. Graphs were shown and discussed—this was, after all, an academic conference, not simply a press-op. He then described some of the hydrocarbon atoms that had in fact been detected, and he qualified that finding: "The reason we are saying that we don't have a definitive detection of Martian organics is that we have to be very careful to make sure that both the carbon and the chlorine are coming from Mars. For example, we have to make sure that the carbon is not coming from any residual terrestrial carbon that we have in our system, or the carbon, for example, could also come from inorganic carbon, the CO_2 you saw released on the earlier graphic. And as the perchlorate [in the soil] breaks up, it might produce, for example, a hydrochloric acid that reacts with some of the carbon and forms these compounds, so there is more work to do." He concluded, "Just one final point . . . in this very exposed material, that is exposed to the harsh environment of Mars, there are many processes that could destroy even the organic material that we expect falls in from space. . . ." He mentioned that cosmic rays, UV, and atmospheric hydrogen peroxide, among others, could toast any organic substances in the soil, "so it will really be an exciting hunt over the course of this mission to find early environments that might be protected from this harsh surface Mars environment and see what we can add to the hydrocarbon and organics story." He concluded his comments and passed the baton to Grotzinger. It did seem that the "mystery" story was all but dead, replaced by a sober and reasoned description of mission progress and possibilities for the future.

When Grotzinger came on screen, he was a bit more somber than I'd seen before—it had been a tough month. I would imagine that his mind had a number of word traps in place; true and honest enthusiasm needed, apparently, to be expressed with some care. It is indeed lonely at the top.

"Okay, I just want to reiterate something that Paul just concluded with, that the instrument SAM is working perfectly well, and it has made this detection of organic compounds, simple organic compounds, we just simply don't know if they are indigenous to Mars or not. It's going to take us some time to work through that, I know that there is a lot of interest in that . . ."You can say that again! "But the point is that Curiosity's middle name is patience, and we all need to have a healthy dose of that, and the reason why is—I'm going to come back to our soil for a minute and try to give an example of that.

"We had to do a lot of work to make sure that this [sample] was some garden-variety, typical Martian soil. We didn't want something that was adventurous, because if we thought that was the case—based on our preliminary assessment using the APXS and the ChemCam instruments—if we thought we had something that was going to be chemically very difficult to work with, we probably would not have immediately put it into the machine. Instead we went through a very long set of triage experiments to make sure that this material would not undergo a state change and maybe evolve water or something while it was in the rover, so we were very careful, and this took about a week or ten days to work through before we could do the first analysis." In short, they wanted a vanilla-plain sample generally representative of most Martian soil that would not do anything unduly weird.

"So being hopeful that there was no gunk that we were passing into the rover, we went ahead to the next step, and what's interesting is that if there is one—if I can try to capture everything you have heard here as simply as possible, what we have is a globally representative material, on Mars, that turns out to also be a rich repository of environmental process and history. And that's what we are trying to do with this mission as we go about assessing habitable environments.

"Every day, we turn on an instrument . . . but you don't really know if it's going to work until you have actually done a measurement. And then once you have done a measurement, you wonder how well it has done compared to all the calibration and baseline testing you have done before you launched the spacecraft. So we go through that each day, and as we turn these on, as one of our team members from Texas decided to call them, we have a hootin' and hollerin' moment, and everybody is jumping up and down in the science team and we get excited about that. But in the end what happens with the SAM instrument in particular is this: SAM comes last. It's at the tail end of the sample-processing chain and is an extremely complicated instrument—it's practically its own mission. When it works for the first time, we have a hootin' and hollerin' moment. But when it works for the second time, you get something that all scientists live by, which is a repeat analysis. You see that what you saw the first time is probably not going to go away, and then, when you do the third sample and the configuration is pretty much the same as it was the first time, you believe that maybe this might just be one for the history books." There it was—a clear explanation of what the statement had been intended to mean, of his mind-set at the time he uttered those ultimately (and unwittingly) incendiary words. It made sense. "That this [result] will stand the test of time as a legitimate analysis of the surface of Mars. That's where we were with the excitement by the science team. The nature of scientific discovery, especially in this business is also very important: we live by multiple working hypotheses. As Paul mentioned, even though his instrument detected organic compounds, we have to first demonstrate that they are indigenous to Mars, then, after that, we can question whether they represent the background fall of cosmic materials that are organic in composition and fall onto the surface of every terrestrial planet," these would be things like the carbonaceous chondrite meteors described in a previous chapter, "then, after that, we can get into the more complex question of whether or not this might be some kind of biological material. But that is way down the road for us . . .

"Finally, serendipity. As any one of us who has worked on Earth understands, on a planet teeming with life, you can look at rocks that are billions of years old, and the probability of finding something that is a sign of life, or even something as simple as an organic material, those discoveries are so rare that every time we find one, it makes it into *Science* and *Nature* [two academic, peer-reviewed journals]. Every new occurrence is a major discovery. So we have to take our time, and it is going to take a bit of luck, and it is serendipity because we are going to think it through well ahead of time and go about this exploration in the most intelligent way that we can, using all of our instruments. This mission is about integrated science, there is not going to be one single moment where we all stand up and on the basis of a single measurement have a hallelujah moment. What it is going to take is everything that you heard from my colleagues and all the other PI's [Principal Investigators, the researchers whose experiments were selected to fly on Curiosity] and their instruments, we are going to pull it all together, take our time, and then if we have found something significant we will be happy to report that."

And that was that. There was a Q&A session, but Grotzinger had summed it up. We have found something that is interesting and *might* be important. *If* it does not turn out to be something that hitched a ride from Earth and *if* the measurement can be repeated, and *if* it does not turn out to be something that fell from the sky as opposed to something of true Martian origin, it will be *very* interesting. If the above conditions hold true, we would then have to try to figure out whether or not the measured organic material is of biological origin. A lot of conditions need to be satisfied, by a lot of people working on a lot of instruments, with repeatability, to get a definitive answer. If that *does* happen, we will let you know. But it takes time, it takes precision, and it takes patience. So please try to embrace the latter.

The "Great Martian Mystery" was over, the press walked—if they had been responsible in their reportage—or slunk, if they had not—back to file their stories. It was interesting to watch the responses. Some outlets simply reported the story of the day's proceedings as if they themselves had never been a part of the frenzy, others related it to the previous week's ride, and still others made it all sound like NASA's fault. The media can be an ungrateful and mean-spirited bunch, Fortunately the last group was small and comprised of mostly lesser outlets.

But there would be more media quicksand ahead for Curiosity.

CHAPTER 25

INTERLUDE: POST AGU CONFERENCE

PRESS RELEASE
12.03.2012
Source: Jet Propulsion Laboratory

NASA Mars Rover Fully Analyzes
First Martian Soil Samples

PASADENA, Calif. —NASA's Mars Curiosity rover has used its full array of instruments to analyze Martian soil for the first time, and found a complex chemistry within the Martian soil.

Water and sulfur and chlorine-containing substances, among other ingredients, showed up in samples Curiosity's arm delivered to an analytical laboratory inside the rover.

Detection of the substances during this early phase of the mission demonstrates the laboratory's capability to analyze diverse soil and rock samples over the next two years. Scientists also have been verifying the capabilities of the rover's instruments.

Curiosity is the first Mars rover able to scoop soil into analytical instruments. The specific soil sample came from a drift of windblown dust and sand called "Rocknest." The site lies in a relatively flat part of Gale Crater still miles away from the rover's main destination on the slope of a mountain called Mount Sharp. The rover's laboratory includes the Sample Analysis at Mars (SAM) suite and the Chemistry and Mineralogy (CheMin) instrument. SAM used three methods to analyze gases given off from the dusty sand when it was heated in a tiny oven. One class of substances SAM checks for is organic compounds—carbon-containing chemicals that can be ingredients for life.

"We have no definitive detection of Martian organics at this point, but we will keep looking in the diverse environments of Gale Crater," said SAM Principal Investigator Paul Mahaffy of NASA's Goddard Space Flight Center in Greenbelt, Md.

Curiosity's APXS instrument and the Mars Hand Lens Imager (MAHLI) camera on the rover's arm confirmed Rocknest has chemical-element composition and textural appearance similar to sites visited by earlier NASA Mars rovers Pathfinder, Spirit and Opportunity.

Curiosity's team selected Rocknest as the first scooping site because it has fine sand particles suited for scrubbing interior surfaces of the arm's sample-handling chambers. Sand was vibrated inside the chambers to remove residue from Earth. MAHLI close-up images of Rocknest show a dust-coated crust one or two sand grains thick, covering dark, finer sand.

"Active drifts on Mars look darker on the surface," said MAHLI Principal Investigator Ken Edgett, of Malin Space Science Systems in San Diego. "This is an older drift that has had time to be inactive, letting the crust form and dust accumulate on it."

CheMin's examination of Rocknest samples found the composition is about half common volcanic minerals and half noncrystalline materials such as glass. SAM added information about ingredients present in much lower concentrations and about ratios of isotopes. Isotopes are different forms of the same element and can provide clues about environmental changes. The water seen by SAM does not mean the drift was wet. Water molecules bound to grains of sand or dust are not unusual, but the quantity seen was higher than anticipated.

SAM tentatively identified the oxygen and chlorine compound perchlorate. This is a reactive chemical previously found in arctic Martian soil by NASA's Phoenix Lander. Reactions with other chemicals heated in SAM formed chlorinated methane compounds—one-carbon organics that were detected by the instrument. The chlorine is of Martian origin, but it is possible the carbon may be of Earth origin, carried by Curiosity and detected by SAM's high sensitivity design.

"We used almost every part of our science payload examining this drift," said Curiosity Project Scientist John Grotzinger of the California Institute of Technology in Pasadena. "The synergies of the instruments and richness of the data sets give us great promise for using them at the mission's main science destination on Mount Sharp."

NASA's Mars Science Laboratory Project is using Curiosity to assess whether areas inside Gale Crater ever offered a habitable environment for microbes. NASA's Jet Propulsion Laboratory in Pasadena, a division of Caltech, manages the project for NASA's Science Mission Directorate in Washington, and built Curiosity.

For more information about Curiosity and other Mars missions, visit: http://www.nasa.gov/mars.

Fig. 25.1. FROM THE PRESS CONFERENCE: The Mars Hand Lens Imager (MAHLI) on NASA's Mars rover Curiosity acquired close-up views of sands in the "Rocknest" wind drift to document the nature of the material that the rover scooped, sieved, and delivered to the Chemistry and Mineralogy Experiment (CheMin) and the Sample Analysis at Mars (SAM) in October and November 2012. *Image from NASA/ JPL-Caltech/MSSS.*

CHAPTER 26

THE YELLOW BRICK ROAD

When people think about driving a Mars rover, some see an image of a scruffy young fellow joysticking to a video-screen spread of Mars as speed-metal plays at blistering volume through his Dre Beats. Others see Big Science (cue the reverb) at a mission control center with endless consoles, blinky lights from a rerun of *Lost in Space*, and white lab coats (for the Millennials among you, I don't have an equivalent, because thankfully these kinds of goofy props went away in the 1990s . . . except for maybe *Doctor Who* sets). But, as we have seen, neither is quite how it works. The delay from Mars at the best of times is profound and averages about twelve minutes each way. Real-time driving is an exercise in patience, some frustration, and slow progress. That's why Curiosity has a high degree of autonomy and some very sophisticated software and computing capabilities that allow it to plot its own course, avoid obstacles, and even begin science operations once it is near a target.

However, there is still an enormous human component. As we learned during the tactical-planning meeting we sat in on, lots of thinking and discussion must take place before the smallest move, and an almost-unbelievable attention to detail precedes every action. A seemingly tiny item, once overlooked, can cause a cascade of consequences, and nobody wants to be responsible for that. So there will be no real-time bumper-car driving, thank you.

So who drives the rover? What does it entail? What about the robotic arm? How do you tell it where to go and what to do? All fine questions, and I had the good fortune to talk to a couple of the stars in JPL's driving and operations team to get the answers.

Brian Cooper is one of the originals—it's said he has the first Martian driver's license, since he drove the Sojourner rover during the Pathfinder mission. He's in his forties, is married, and has one daughter whose second middle name is

Sojourner. Now that's company spirit. He has been at the lab a very long time, and in his spare time he studies computer gaming code and plays the games as well (you gotta be well-rounded, after all). We talked in front of a computer that he uses to write program code to enable Curiosity's drives across Mars.

"I was the first rover driver on Sojourner, so I guess I'm one of the originals," he cracks. "I also worked on MER, Spirit and Opportunity. That was great. At some point I stopped for a couple of years to develop this tool we are looking at, that is, to modify the tools we are using to drive the rover. It's called RSVP, which stands for Rover Sequencing and Visualization Program."

Fig. 26.1. RSVP: It's not a party invite—it's software that Curiosity's drivers use to program the rover's travels across Mars. In this image, the rover is superimposed over a photo of the nearby terrain, with lines leading back from its wheels to indicate the track from where it came. *Image from NASA/JPL-Caltech.*

I ask him what a typical day is like for a rover driver: "Well, we have this notion of a yellow brick road, if you will—where we are and where we want to go. What I look at every day is these two screens. One is a screen that has commands that we send the rover—we can type in commands and values that get turned into binary bits and are sent up to the rover. This is where they happen . . ." He points out computer software that looks like something from the Windows 3.1 era but is in fact written in modern Unix. It is not fancy looking because there is no reason for it to be. "And over here is [a] dictionary that has everything that we can tell the rover to do. There are many thousands of commands."

I look over the scrolling list that would probably stretch out to the parking lot if printed. How the heck do they keep track of what command to use? "We use this search box here"—that should have been obvious, I suppose—"and it defines certain classes of commands, for instance, mobility. You can also constrain things you are seeking." It still looks complicated, but then again, I have no background to judge by. Is it a tough learn for the newbies? "Well, before anyone can sit in the seat, they have had a *lot* of training. So it's not often that you come upon a command that you haven't used before."

As the programmers come up with new commands for new needs, they get added to the huge list. It's a pretty elegant and almost-simple setup once you understand what it is designed to do. Of course, just because it appears to be straightforward from the operator's point of view does not mean that the process is not immensely complex.

I ask him to show me how they prepare for the daily drive, usually scheduled for the following day: "Well, what we have here is a 3D representation of the rover as it is right now." He points to a virtual rendering of the rover sitting on the part of Gale Crater it is currently driving across. You can zoom in or out, see an overhead view, look from the POV of the rover, and much more. It is very cool. "If you look here you see the tracks that the rover made on the ground during yesterday's drive." I ask how they understand the true shape of the surface. "These surface maps are created by taking images from various cameras on the rover, which are provided to us in stereo pairs, and then you can run machine algorithms on them to determine the depth of the terrain." This is done by evaluating color and shadow with complex software designed specifically for this task. "Then, once you have that, you build up a mesh and then colorize it. There's also orbital data in there. This gives us the view from up close and then we can switch to other views that we might've loaded." OK, got it.

It is not surprising when I later find out that he has experience in, and a great passion for, computer gaming. The software is not dissimilar. "We're certainly not state-of-the-art compared the latest game engines," he says. "But I've always had an interest and expertise in computer graphics, and I like games, so I've followed that research over the years. I tend to think in terms of imagery and spatial relations, so for me this is intrinsically fun." He points back to a command-line-driven software he showed me earlier. "On the other hand, the author of the RoSE [for Rover Sequence Editor: it controls the graphic user interface or GUI] program—right here—is a dual computer science and English major. He's not fascinated by the 3D imagery, he's fascinated by *words*. He likes prose and syntax, and what he has created is very elegant." It truly takes all kinds, and not everyone here started as a hard-core space guy. I'd guess that roughly half of the people I've met at JPL felt destined to do this since childhood; the rest came to it somewhere late in college or thereafter.

He begins to plot out tomorrow's drive as I watch. "This software is an improved version of what we used on MER. For instance, having this terrain model from the HIRISE orbiter camera is a new feature." HIRISE is the high-resolution camera on the Mars Reconnaissance Orbiter. "We didn't have that on MER, so their maps had a flat plane instead of a 3D model. So this improvement is very useful, as it tells us where we've been and where we are going, and things we want to avoid, like craters. I can exaggerate the terrain up to five times, which is kind of wacky, but if we are not quite sure of what the terrain ahead is like you can exaggerate it and instantly see certain features." He moves the mouse, and the terrain suddenly bulges from normal-looking to some kind of psychedelic cartoon—I wonder if this is what my classmates in high school who spent so much time in semicatatonic drugged states experienced. It would explain a few things. "Then you can see where the hazards are. So the shading is based on whichever camera they're using and the time of day and lighting and so forth." It is truly amazing software, given that it is utilizing data from an orbiter a couple hundred miles up and from small, ground-based cameras on the rover.

But just knowing—and being able to see—the terrain is not really enough. Rovers are very sensitive to the slopes and the angles the landscape imparts on the machine, so the drivers also need to have an idea of how the rover will interact with it. I ask how this is accomplished. It took a bit of time to explain to me, but I'll do my best to condense it. It's cool and worth it, so bear with me.

Brian changes views on the screen, and what is up now is a wider view of the 3D model of the terrain ahead of Curiosity. The virtual rover is no longer in view and has been replaced by a little red cone, which he points to. "This little 3D cone is called the *point normal cursor*. So you turn it on, and when you hit this feature"—he performs a lightning-fast keystroke—"you can drag the cursor around the land-scape, and it actually hugs the terrain." It looks like an inverted traffic cone, as he drags it from point to point, and wherever he "drops" it, the cone magnetically snaps to the nearest surface feature. "If there is a rock or an incline, it will tilt. It's defining the normal to the surface."Wherever the cone gets placed, it deviates from vertical to indicate the severity of the slope it is placed on. "The important thing is that we can get a sense of where the rover is and what the terrain around it is like, and whether it is safe to drive. So I can move it around . . . notice how it's changing its tilt as I move? We're going up little slope here. It's kind of like being a kid with a Tonka truck. This is very useful in particular for planning arm motion so we can place the arm sensors right on these targets." He can define a spot where the cone has been and drop what looks like a smartphone app pushpin. "It's another way that we communicate between the science team and the rover planners."

As he demonstrates the software, I notice that there are grayed-out areas, regions with no detail. "Those are areas of which we have no knowledge. It's because when the image was taken from this point of view it was obstructed." So previously, a rock or rise in the ground might have blocked the Mastcam's view. "We couldn't see what was behind the obstruction. That's something you have to take into consideration—do you want drive in that area, with no knowledge or it, or do you go around? The answer is usually to go around unless you can use data from other sensors."

Once again, safety first.

So how is all this data and the software actually used? In general, they want the rover to drive itself autonomously when it's safe for it to do so. Between the long delays of commands going up and results coming back down, as well as the limited alignments of the relay orbiters, real-time joystick driving would be horribly slow. Better to tell it to just drive on its own, based on parameters and goals entered into the RSVP software.

"We can tell the rover to do autonomous driving. That's how we drive long distances. We typically do what's called a blind drive up to forty or sixty meters, depending on how flat the area is. Then we will tack on an extra forty or so meters. Say we are trying to do a one-hundred-meter drive for the day. The first forty might be in autonomous mode, where I'm telling it to go here"—he points to a rock on the map—"but I don't care how you get there. We've already programmed in things that are considered hazards, for instance, what's considered a rock to avoid and things of that nature. The downside is it takes longer for the rover to do that because it will take a short step, say, a meter, take more images, process them on board, and build an internal map. And all this occurs while it is stopped."The rover, once it has received instructions and a basic map from JPL, needs to refresh that map as it moves. So the sequence is: drive, stop, shoot pictures, insert images into the software, interpret them, then verify the course ahead or generate a new one before driving farther. Rinse and repeat. And it's amazingly smart software for the small and comparatively slow computer Curiosity carries on board. "If you put Curiosity in a maze, it can actually find its way out by using this algorithm called D-STAR. It knows how not to get stuck in a cul-de-sac. That's how we can let it drive beyond the areas where we have high confidence. It often knows better than we do." I could have used this thing's brain the last time I parked at Disneyland.

Curiosity is currently crossing a flat plain, but progress has slowed due to the risk of wheel damage: "It's gotten to be nerve-racking in the last few weeks because we've noticed that we were doing damage to the wheels. We don't want to do that any more than we have to. We hadn't really considered that up until this point, we said 'Well, we don't care too much about little rocks, we can drive over them.' Of course we didn't want to go over a cliff or be stuck in the sand dune or a sand-filled crater. Those are the kinds of things we've been avoiding.

"But Mars throws monkey wrenches at you all the time." Hmmm . . . seems like I've been hearing variations on this quite a lot. "Now we have to learn how to drive again, and we are starting over with shorter steps. We're trying to get out of this area and then we will have a lot fewer pointy rocks and we can make faster progress." A few weeks after I spoke with him, Curiosity made a detour across a wide sand dune—a nerve-racking feat in itself—to get out of this wheel-killing terrain. To everyone's relief, the rover did not bog down.

Within the year, Curiosity will be making its way up into the foothills of Mount Sharp. Will operations change then? "As we get into the foothills and begin to go steeper and steeper, and then especially when we're doing arm operations, it's going to be challenging because there are limitations to what you do with the arm in various height and tilt regimes. Specifically, drilling and sample transfer and things like that."

Arm operations is another specialized skill not found in your average video game. Vandi Tompkins is a rock-climbing, airplane-piloting adventuress who could give Lara Croft a run for her money, but fortunately for the MSL mission and us, she chose to be a rover driver, arm operator, and engineer. Coming from a fairly traditional home in India, accomplishing this was not a sure thing—there was an expectation from her family that she might pursue a more traditional path—marriage, kids, home, and hearth. But once you meet her, you can understand that most people, her parents possibly included, would make a suggestion and then step aside, for Tompkins is a determined person.

Fig. 26.2. A NEW GENERATION: Vandi Tompkins, rover driver and programmer, poses with her previous "ride," a twin of the MER rover. *Image from NASA/JPL-Caltech.*

With a doctorate in robotics from Carnegie-Mellon, her first job with NASA was at the Ames Research Center in Northern California. Before too long she had relocated to JPL and was working with the Mars Exploration Rovers. What else would someone with a degree in space robotics want to do? Her husband of just a few years works over at SpaceX, Elon Musk's private rocket company in Hawthorne, about twenty-eight miles across LA from here.

"I do three primary roles on MSL, and as a rover driver we tend to have three specializations: driving or mobility, operating the robotic arm, and the sampling system. Each of these is such a rich domain. There are a couple of us [who] do all three right now and eventually they would like everyone to be able to do that. There are fourteen drivers as of now."

So what is involved in "driving" the arm? How hard can it be? Pretty hard, as it turns out. "It's [a] five-degrees-of-freedom arm. There are all these instruments on the end of the arm: the drill, the CHIMRA instrument, and so forth. It's pretty heavy. So we have done a lot of testing related to that . . . trying to break it in various ways, all to make sure that that it will work on Mars." It turns out that with a metal pole as long as the arm is, with joints and motors and heaters and so forth, there are a lot of variables: how much will it sag when extended? How much power will you need to move the arm? To run the devices on the turret? To keep it heated enough to not freeze up in the cold Martian night?

Tompkins continues: "What makes space robotics so different is that you don't have many second chances. You really have to dot all the *I*'s and cross all the *T*'s, so to speak. I'm interested in having the rover in situations that are not the nominal working situation." This woman loves a challenge. "This giant, hundred-kilogram arm will drop a baby-aspirin-sized sample very precisely in these inlets and the instruments. An instrument like SAM, which is so capable, needs clean samples. We have to make sure the sample is not contaminated. Our general mode of operation is that instead of doing something risky, we want to have the time to stop and wait and call home. But there are times when we have to do something autonomously and that is also very interesting."

Interesting is the term engineers often use for *challenging* and *potentially risky* operations. You have to love that.

Nothing is easy on Mars, and there are a lot of robotic acrobatics the arm must do to move a sample from one chamber to another inside the CHIMRA—a soil-holding and size-sorting chamber on the turret—and sift out the fine particles appropriate for the instruments inside the rover. "You have to very precisely position the arm because you have very close clearances. Once the arm is in position, you open the inlet [that leads to SAM and CheMin] just when you need to, and, for example, the SAM instrument has carousels of little cups built in. It places the equivalent of a little trash can under the opening, because when we bring the arm over, we're afraid there might be things on the outside of the turret that might fall, and we don't want to have the sample contaminated by them. So, once we get the arm into position, we move that trash can out of the way and then get the collection cup under it. Then you drop off the sample by vibrating the collection device. You then have to move the arm out of the way to an intermediate position because you can't yet close the cover on the inlet to the instrument." The tolerances and distances are so close that you have a tension between getting the tiny sample into the cup without losing it and not banging into the rover. "It has to be close because wind and other weather effects could blow away the sample."

On the Mars Phoenix mission in 2008, which landed near the Martian north pole and collected icy samples there, they discovered just how trying this could be. Remember that there is a long delay in two-way communications. When controllers positioned that spacecraft's arm above the receiving funnel and tried to drop the sample, prevailing winds kept blowing much of the sample away. They learned a lot in that mission.

There are other risks involved with arm operation and the instruments on the turret. "There can be a problem with the brushes," she adds as an example of this. There are wire brushes on the arm that can be used to clean a rock before it's sampled—that's the DRT tool I discussed earlier. "If you leave them [the wire bristles] in contact with the rock for too long, they can get bent, so that's another example of a case where the arm will autonomously move itself just enough so that is not in contact. I think it's hugely interesting." There's that word again. Demanding, maddening, complex . . . *interesting*. I envy the passion evident here.

"When the arm is making a sample deposit, it is software commanded, but it also has to be compensated because, for instance, the rover can be sitting on different tilts or inclines, so we have to do various moves to make the drop-offs work. That's something we simulate on the ground, then we expect it to do the same thing on Mars. We have a simulation which includes the rover's attitude so we know where it's going to go. We have to test all the possible situations and conditions under which the operation could be interrupted."

At this point, I am a bit confused. There are software simulations of how the arm will operate in certain conditions, but there is also a physical clone of Curiosity at JPL that can be used to test arm motions as well. How do they decide when to use software, and when do they resort to testing with the physical arm on Earth? "For everything we have to do the first time, for instance, the first three months when we were working on Mars Time, we had many days like that . . ." that is to say, many days during which they needed to use both simulation modes—software and the physical twin of Curiosity—to test some of the arm's activities before sending up instructions to Mars. "But now that we have been on the ground for a while, some activities are repeats—we've done that before, and we really have very high-fidelity simulations. So, for example, let's say that the science team wants to get a MAHLI image of the drill hole, which is something we did recently. That is something we feel very comfortable doing because even though we haven't done that *precise* activity before, we have run it in simulation, and that's where experience and understanding comes into the picture." When they drilled into Mars, they took close-up pictures of the drill hole from four angles to see exactly what it looked like all around its circumference.

So how do you translate a general simulated activity into a specific situation? "The Navcam will take a stereo image, and that gives us depth, and from that we can create a mesh." It allows them to add that into the software simulator. "Now there's a certain amount of error in that, and the gyros can drift," that is, the inertial measuring sensors—the gyroscopes—can incur some drift and show an erroneous location of the arm or even the rover itself; we are talking tiny fractions of an inch here, but that's enough to be a big problem. "There are also things you have to take into account regarding position accuracy of the arm. You have to have a good understanding of all that so you know how close you are getting to something. But we feel very comfortable with the understandings we've gotten from the simulations."

I ask for an example. "Okay, . . . recently we wanted to dig a trench. We've done this with the other rovers. You spin one of the wheels around to dig a trench. That reveals the sub-layers, which have a more interesting composition. It's not been exposed to the surface and the elements. And then you want to be able to do contact science," for example, to use the APXS or scoop or drill right at the rock or soil exposed by creating the trench. "Now our turret is so mechanism rich, it's very crowded; the clearances are very close. So if you want to use the APXS, you have to be careful where the drill is and also where the target rock is. When it's flat, it's not such an issue, but when the rover is in any kind of terrain with relief, it is." Complicated, that is. "So in cases like that, we will simulate them in the test bed," with the Earth-based rover twin. "In that case, you have to take into account that it's operating in Earth gravity. For example, with the motions we did to dig a trench, we know that because of the weight of the rover here on Earth we will go deeper . . ." In this case, they would have to take this into account and change the variables. "Part of the decision is a judgment of what do I need to use in this the particular situation, is it sufficient for me to have tested it in simulation? Or do I need to use the test bed?"

But it's even more complicated than that. "Another example is sample flow. Samples behave differently on Mars, so we have another test bed in a chamber that simulates some of those conditions with the rover in it. Part of the problem is figuring out which level of simulation we need to use. Over the years, we've learned a lot. Many of us also worked on MER, and some of us went to graduate school for robotics, as I did, and you have experience you bring in from that."

"It's amazing how much we know about Mars now. Our team's approach works really well. To have people [whom] you can talk to, like soil experts, is very helpful. They can tell you what the soil is going to be like underneath, for example."

So that's how you drive a rover, how you teach a rover to drive itself, and how you operate the robotic arm. It's complicated, and much of it has to be software driven and autonomous; there's just too much delay to do it in real time. Pathfinder blazed new trails with the slight autonomy that Sojourner had. With the MER rovers, ten more years of development and learning have been a huge benefit. Now, with Curiosity, all the hard work has paid off.

"I think we work very hard to try to eliminate any possibilities of failure," says Tompkins "But you can't eliminate it entirely—it's a matter of probability, and you can't get it down to zero. So it's really interesting to work on development and testing. You like to think you've done everything you could, but there are still things that [you] could've done even more work on. So when it works, and it works better than expectations, you really realize that each day is still a new day, that anything can happen. The Martian environment is really harsh." And probably . . . *interesting*.

If Opportunity's longevity is any indication, there could be a decade or more of exploration and discovery in store for Curiosity. And, being such an incredibly capable and advanced machine, every month of operation builds on MER's rich history and adds immense amounts of data to the books.

And then, of course, there is Curiosity's huge increase in capability for evaluating and gathering samples never available before. And that brings us to the long-dreamed-for ability to find materials not exposed to the harsh environment of Mars . . .

That is the magic of the drill.

CHAPTER 27

THE DRILL

There are a few things that set Curiosity well ahead of previous rovers.

It was bigger.

It was heavier.

It had a built-in lab to analyze soil and rock and atmospheric samples. In fact, it had 180 pounds of scientific instruments aboard its roughly one-ton chassis; the MER rovers had fifteen pounds of far less capable ones.

And, perhaps most notably, it had a *drill*.

This last item was just as remarkable as the onboard laboratory and almost as hard to engineer. I've talked about the science instruments—SAM and CheMin are fantastic machines—and discussed the scoop and the delivery system for Mars soil earlier, but I've been saving the drill until now because it's more interesting to hear about it when it is about to go into a rock.

A drill had been one of those Holy Grail, wish-we-could-have-that items on the planetary scientist's packing list since Viking. On that mission, if they wanted to investigate rocks or dirt that had not been bleached out and sterilized by the sun, they had to use the scoop on the end of the arm to shove things around as best they could. Viking was a static lander, so there was only so much that could be reached with the arm.

With Pathfinder's Sojourner, it was a different story, but with a similar outcome. The little rover was mobile and could travel an area tens of times that of Viking's reach, but Sojourner was the size of a toaster oven and barely outweighed many of the rocks it was looking at. So when JPL wanted a "fresh" sample to look at, they shoved a small rock over or dug a shallow trench (more like skimming off the surface dirt, really) by locking five wheels and spinning the sixth. It was better than nothing.

The MER rovers, Spirit and Opportunity, were more robust, weighing almost four hundred pounds each. They had robotic arms, more mass and larger wheels. Still, the best they could manage would be to flip some rocks, dig some trenches, and scrape off some surface patina on the rocks with a spinning wire brush. The technology showed progress, but, ingenious though the creators and operators were, the MER rovers had their limits.

The Mars Phoenix lander, which set down near the Martian north polar region in 2008, was also able to use a robotic arm to scoop and scrape through the icy crust up there, but it was at the mercy of soil variables, including hard ice, as well as the short span of the mission (about five months).

Then along came Curiosity. The geologists wanted a drill, which was a worthy goal. After the MER rovers and Phoenix, it was understood that the surface of Mars was a far-nastier place than had been suspected since the Vikings landed there. It was baked with solar radiation/UV and cosmic rays. There was bleach-like perchlorate in the soil, and the red hue of the sand and dirt and rocks was a product of billions of years of oxidation. In short, the surface was a sterile, radiation-blasted, bleached-out desert. It was possible that all of Mars, above the surface and below, was like that . . . but we would never really know until we could see *inside* a rock. And the only way to do that was to break one with a mechanical hammer or to drill into it.

In the end, the engineers really built a bit of both.

The drill on Curiosity is a masterpiece of ingenuity. It's a percussive drill, meaning that in addition to spinning like the hand drills we are all familiar with, it also hammers on the rocks as it spins. Just that mechanism is difficult enough to build to make it strong and reliable enough for use 150-million-plus miles away, but there is more. It also needed to *collect* the sample it drilled. Somehow, the powder that resulted from beating, chiseling, and grinding the rock had to make its way to the labs on board Curiosity. That's a challenge.

Oh, and one more thing. The rover had to carry more than one drill bit. Curiosity was designed to fly to Mars with the drill bits in a sterile box and the chuck of the drill empty, but as you will recall, a last-minute decision (and one that stirred much controversy within and outside of NASA) was made to place a bit in the chuck before flight, just to make sure that there was at least one ready to go. But if that drill bit broke, or got stuck in a rock, or just plain wore out (it would be drilling rocks, after all, and they tend to be hard), the engineers wanted to be able to have spares on board. So a rather-elaborate system was designed that allowed the drill head to let go of the worn, stuck, or broken drill bit and reach down and clamp onto another.

Whew.

So how do you build and test such a device? As with so much related to this mission, the process of designing and perfecting the drill for Curiosity could fill its own book. But we'll settle for part of a chapter.

In the broadest terms, the drill bit is attached to a chuck, or receiver, that is in turn attached to a mechanism that both twists (to drill the rock) and percusses (to beat the rock). So the drill assembly is attached to a thin metal diaphragm. There is a magnetic coil—much like a voice coil (for those who know how speakers or many microphones are made) and the diaphragm has enough "give" that this coil can cause it to flex in and out and cause the drill to hammer as it spins. This, in addition to the drive motor (that makes it spin), creates a drill that twists and hammers at the same time. It can be set to either drill or hammer alone, but it is really designed to do both, as that is the most effective way to drill a rock.

Now, remember that this drill is mounted on the turret on the end of the arm along with the MAHLI camera, the APXS instrument, and other devices, so it can't be allowed to beat things up too badly on the turret while it's running. It was insulated enough to avoid damaging anything nearby. And on top of all of this, it had to be ultrareliable (like everything else) and lightweight enough to fly into space.

The engineers designed, built and tested, and tested some more. As they learned of the limitations of the design and its weaknesses, they reengineered it, rebuilt it, and tested it again.

Late in the game, shortly before launch, in fact, they discovered that there was a problem. The drill could cause a short circuit, which would theoretically take down a good part of the rover's electronics. Nobody I spoke with called it a panic, but it had to come close. Soon after the engineering team had a fix, Rob Manning was working on it, jiggering it in place, while the rocket was actually on the pad waiting to launch. It was that close.

I'm going to let one of the people responsible for this fix action narrate the process. We met Dan Limonadi during our tactical-planning meeting up at JPL; he was the Captain America–type I droned on about: tall, lean, just a hair over forty, and a community icon outside of work. A large part of his responsibility on MSL was the arm and its instrumentation, which includes the drill.

Fig. 27.1. ALL BUSINESS: This image of Curiosity's robotic arm shows the drill (as indicated by the arrow), facing the camera in mid-turret. The bit is a jagged, percussive design. This photograph was taken in Yellowknife Bay. *Image from NASA/JPL-Caltech/MSSS.*

"In the late summer of 2011 we were doing a lot of testing with Curiosity's sampling system, and we had multiple issues that hit us all at once. One of them was a short in the arm and the drill, so we scrambled to see what was going on and it was a very intense time." See? No panic, just *intense*. He did not mention that a lot of twenty-hour days were involved, but they were. "We came up with a quick fix that would allow us to operate in the face of an electrical short should it occur. We had to come up with a design, review it with a bunch of people, and then work it in at the Kennedy Space Center with an already-buttoned-up spacecraft." That is as last-minute as it gets. "That was a very exciting time in [a] not-such-a-good way. We actually discovered the fault on the engineering model [the nonflying twin of Curiosity]. This was a funny twist of fate." Again, with the understatement.

"In the drill assembly there is a spring attached to the voice coil and the percussion drill. This hits the percussive mechanism. The coil uses magnetic force to move the hammer back and forth, so the spring gives it a bit of a natural frequency. It turns out that the spring in the engineering model was a little bit out of true. The team knew that when they put [it] together, they knew they didn't quite have the right spring that they needed to get testing going, but we were short on time. It turns out as we when were test drilling with that spring, the side started rubbing and it caused a failure—one that we would never see in flight with the proper spring. It didn't actually short the wires inside the drill, but it did expose some of the wires." So the failure did not actually occur, but in theory it could—just not on the final, flight version. But you simply cannot be too careful. "Once again, this is something that would never happen in flight with the better parts." All good and true, but it's not enough to let the engineers sleep at night. It had to be fixed.

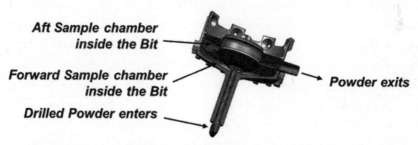

Fig. 27.2. THE DRILL EXPLAINED: This schematic view of Curiosity's drill shows *(from bottom left)* the sheathed drill bit—the sheathing encourages powder from the rock being drilled to crawl up the bit traces into the collection area; the first chamber where rock powder collects after crawling up the bit; the second chamber where powder migrates after being stored in the first chamber; and finally *(on the right)* the exit orifice to send powder to the analysis instrumentation. *Image from NASA/JPL-Caltech.*

Limonadi continues: "But what it did do was expose the *class* of fault just in time to do something about it before takeoff. It was a minor miracle and a bit of fate, good and bad timing. It was funny in retrospect that [it] happened when it did. So we might [have] been in a different situation if we had [to] operate that drill without having had a chance to put in some mitigation."

"As we dug into this, we started testing other units more rigorously, and they failed in a different way, but with the same result of shorting out the chassis. So we learned a lot of different things that we had fix, to do proper mitigation. We discovered that we could actually induce a short in the arm, sending current into the wrong place by choice, and detect every bit of rattling so that if need be we could turn off the drill before it gets damaged."

In short (no pun intended, I swear), they did a couple of things to fix the problem. First, they made very sure that the spring on the flight unit was perfect and would not cause the first problem they had detected. Then Manning and his team designed devices to better secure anything else that could cause a short in the drill assembly. They also figured out how to ensure that if they *did* see any issues during operations on Mars, they would be able to shut the drill off before it could hurt anything else. There was also some discussion of how to route the electrical current to avoid any damage in the unlikely event of a short, but that is above my pay grade and likely above your level of interest. Suffice it to say it was fixed.

It is worth mentioning that the drill tests that exposed the short were far beyond the expected use of the drill over a period of years. We are talking literally tens of hours across many weeks of use at the top setting. This is way more drill time than the entire primary mission of one Martian year (two Earth years) will utilize. But that is how these guys roll—test for maximum effect, simulate the worst-case scenario, and fix everything possible. Now you know why Opportunity has been roving Mars for over a decade—this level of dedication pays off.

The expectations for drilling activities were also downgraded about this time, so the demands were lessened. This changed the target run time for the drill. Ashwin Vasavada remembers when NASA changed the number of expected drill uses for the mission. It had been a number that, as drill testing proceeded, was starting to make a lot of people uncomfortable. It was up in the seventy-hour-plus range. Then—"NASA changed that requirement for us to twenty-eight [hours]. When we actually mapped out how many calendar days there are in the two-year mission, how long does it take to get through all the events we had to accomplish, twenty-eight was the right number in terms of the acquisition for the drill. So that's what the system had to be designed to do." Whew—a break at last. The drill would only be expected to function for about a third of the original time specified. But even this new number—twenty-eight hours—was a lot for a mechanical device as violent as a percussive drill. "You're talking about twenty-eight times you're going to put a drill into a rock. The rover could slip and the drill bit could get stuck, all those concerns. At that point, you don't know what the rock will be made out of, it could turn into goo and clog everything up." And that's just grabbing the sample. Then you need to investigate it. "Then you need to run those twenty-eight samples, and that means putting them in a mass spectrometer with [one] part per billion accuracy! Talk about not having cross-contaminations between samples. For a long time, I don't think that anybody thought it would be possible. It's not even just drilling robotically—with a science sample, even if you did that in a university lab setting, people would be wearing a gown and gloves and all that. So that's what we had to invent." It was their own little biohazard-level-4-caliber roving clean lab.

And there was still another bug in the ointment: Teflon.

Vasavada explains: "We also had a Teflon contamination issue. This came up in the same timeframe as everything else, between September and November of 2011. This is a really challenging issue because you have a complex system with a lot of moving parts designed to break the rock you are trying to sample. It's also trying to hold onto and collect the powder the drill creates. So there is a can, two chambers, in the front of the drill, right next to the bit, to collect the powder. To do this, there's an exit perpendicular to the drill."

It helps to understand that a tiny bit of rock powder was supposed to crawl up the spiral on the drill bit and land in the collection chambers—the CHIMRA unit—but it was impossible to know what angle the entire assembly would be in when it was drilling. It might be drilling straight down, or to the side, or any angle in between. "We were supposed to be able drill anywhere from vertical to horizontal, and remember that the robotic arm has five degrees of freedom—a human arm has seven degrees of freedom by comparison. So if a science team member wants to place the drill against a horizontal target, we can't control the orientation of the little exit tube." The drill could, in theory, be in almost any orientation when in operation, except possibly upside down. And no matter what angle it was operating from, the powder it generated needed to travel up the bit, collect in a chamber, and stay there long enough to be processed.

The answer was to have two collection chambers so that the second chamber would collect the powder from the first before it spilled back out. It would take a bit of clever arm choreography and a lot of testing to make sure that it worked in all possible orientations, but they figured it out.

But what about the Teflon? Dan Limonadi takes the story back up: "Recall that this is a percussive drill and you need get the hammer force to the bit. Around that bit at the top, these chambers need to have a flexible interface which is thin titanium with a small circular clamp holding it; it's a friction fit around the edges. That seal was Teflon-impregnated fiberglass. So the issue we had was, you hit the diaphragm thirty times a second and there is no way to avoid chafing some of that Teflon-impregnated fiberglass off. And the SAM instrument is supersensitive, and you have to make sure that you don't get a mote of dust or something from the sampling system in there that is not supposed to be there. It's almost an impossible job. The tolerances are incredibly small."

SAM measures things at the molecular level, and anything that was not Mars dirt was a no-no and could screw up the readings. The engineers had another talk with the science team as they were trying to design something that seemed to be impossible.

"In the final analysis, it ended up not being that big a deal once we were on Mars. The SAM team concluded that the small amounts of Teflon they might see could be a little annoying, but it's not the end of the world. [They] didn't expect to see much in the sample. So for launch were thinking about the worst-case scenario and wondering if that was going to be a problem. That caused a big stir, but ultimately the SAM people said it was going to be fine."

Also, for the most part, Teflon in the sample could be accounted for when the results came down from SAM anyway. "Teflon is a chlorofluorocarbon, and there are no natural compounds that have that chemistry, so there wasn't really a risk of a false positive. You can tell this is a man-made carbon compound. There was, however, a concern that if there was enough Teflon mixed in, it could possibly mask some of the natural Mars-originated materials found in the samples. If you have ten parts per million of Teflon and one part per million of Mars organics, the Mars organics might be hidden by the Teflon. But in the two drill samples we've done so far, the SAM team says they have not found any Teflon in the sample."

He summed it up in a way that could describe so much in the MSL mission, where any errors or faults could be amplified by an exponent over anything that had gone before. "There was an interesting [there's that word again!] pucker factor, and everybody needs to get together and hold hands before launch, but it turned out to be the right call. It's a testament to how conservative we are at JPL. Once we got to the rocks, they were very soft, so I don't think there'll be an issue within the drill's lifetime."

From your mouth to Mars's ears, my friend.

CHAPTER 28

CALLING JOHN KLEIN

By the end of 2012, Curiosity had descended into Yellowknife Bay proper and was looking for a rock to drill into. Of course, there were plenty around—everywhere you looked, in fact. But it had to be big enough to sit still for the drilling, and preferably the first target should not be too hard—nobody wanted to break the drill on the first try. And, of course, the rock should be positioned in such a way, and of the proper composition, to yield a maximum science return.

Like everything else in this mission, patience was a virtue. John Grotzinger meant what he said after the whole "history books" debacle.

For the press, the news conferences at JPL were mostly over, at least ones that we could attend. Most were now being held by teleconference, which is a great time-saver. JPL and NASA have the system pretty well nailed down, and I was eagerly anticipating the results of the drilling along with tens of thousands of others.

Yellowknife Bay was a shallow depression past Glenelg and Rocknest. It sounds like the rover is really driving over hill and dale to all these places with funny names. But they are not quite what we would normally think of as distinct "places." The distance from Glenelg to Yellowknife Bay is about the same as from your back fence to the front door of your house—something in the neighborhood of two hundred feet. It's a bit like naming the rocks, shrubs, and gopher holes in your yard. But at the speed that Curiosity was driving, and given the extreme and rich geological diversity of the part of Mars it was traversing, it makes sense to create names on a map. For one thing, it helps the drivers and science teams (not to mention the rest of us) keep track of where Curiosity is. It also makes Mars feel friendlier. It's not much fun to move from rock N165 (the numeric designator for Coronation) to 4.59°S by 137.44°E (the positional measurement for Hottah).

As the rover moved into Yellowknife, the geologists started looking for a drilling target. They had lasered rocks, sniffed boulders, and tested the sandy soil of

Rocknest. The atmosphere had been analyzed. The relevant instruments, SAM and CheMin, had been commissioned and tested. The last major item on the mechanical list was the drill.

As the rover made the shallow descent into the "bay" (just a few feet of gradual drop, really), the Mastcam and ChemCam were busy looking at potential targets. The wonder of the laser-driven ChemCam was becoming apparent now; what could have easily taken days or even weeks took only hours. With the ability to shoot the laser and image the resulting spectra from a distance, the device saved not only drive time but also the long and careful process of bringing the arm down and rotating the APXS into place to examine a rock.

One more tool that needed to be tested was the wire brush, the Dust Removal Tool (or DRT). The rotary wire bristle was capable of scraping the dust off of rocks for a better look. This too would help them in the all-important job of finding a rock to drill.

The wire brush was used on a rock they named "Ekwir_1." Diana Trujillo, lead for the brush team (everything on Curiosity has a team) said in a NASA interview, "We wanted to be sure we had an optimal target for the first use." She added, "we need to place the instrument within less than half an inch of the target without putting the hardware at risk." As we heard in a previous chapter, they even obsess about how long and how hard to press the DRT against the rock in order to avoid bending the wires on the brush.

And they even had to take the ambient temperature into consideration. Anything on the arm that actually touched a Mars rock, or even the soil, was at potential risk of breakage, or causing arm damage, due to any forces imparted from the target object through the instrument and into the arm. The rover could always slip, but they were careful in these early efforts to pick flat areas on which to park Curiosity while running tests. But if the temperature shifted too much while they were working, even the small expansion or contraction of the arm could be a problem. Of course, this was factored into the simulator, but let's look at a theoretical example. Say they were using the drill on a rock. Something snags, or the rover has a software hiccup that freezes the arm overnight. The temperature differential between night and day in Gale Crater, at the time, was almost 120°F. That broad temperature swing would, as it got colder, cause the metal in the arm to contract—and much more than you might think. As the arm contracted, if the drill was sitting in a borehole, it could at the least bend or even break the drill bit. At worst, it could destroy the drill apparatus or damage the arm servos. Then, as the sun rose, the arm would heat and expand. The combined forces could make for a bad morning.

So with either the drill or the brush, nobody wanted the arm moving in ways that were not planned for. So they picked a nice, flat rock on a nice, flat surface to work on. Once the DRT instrument had done its work, the MAHLI camera was used to take a closer look. The resulting images were something to warm the hearts of the geologists. The rock went from the usual ruddy, oxidized red to a grayer shade, and pits and veins could be seen. It was not remarkable, but it was a happy day, as the brush worked and they could move on to find a drill site.

They soon had their man, so to speak. A flat, veined rock the geologists named John Klein was selected. The namesake John Klein had been the deputy project manager of MSL until 2009, and died unexpectedly six months before launch. It seemed a fitting tribute. Curiosity slowly rolled to the rock.

Part of what attracted them to Klein and the area in which it dwelled was the oft-cited property of thermal inertia. Orbital images with infrared cameras had shown that this area cooled more slowly in the Martian night than other surrounding terrain. The slower a rock cools, the denser it is. And dense rocks in a lake-bed environment could mean sedimentation, and the denser the sedimentation, the older it might be and the more information it might hold. As Grotzinger put it, "The orbital signal drew us here, but what we found when we arrived has been a great surprise . . . this area had a different type of wet environment than the streambed where we landed, maybe a few different types of wet environments."

Before the scientists finalized their selection of this rock, the rover shot the veins in it with ChemCam. The veins seemed to be something that had been wet in the past, and had the general makeup of gypsum here on Earth. Richard Cook, the MSL project manager, said in a NASA interview, "Drilling into a rock to collect a sample will be this mission's most challenging activity since the landing. It has never been done on Mars. The drill hardware interacts energetically with Martian material we don't control. We won't be surprised if some steps in the process don't go exactly as planned the first time."

Once Curiosity got close to John Klein, the DRT instrument was used to clean off a portion of the rock and Ken Edgett's MAHLI magnifying camera could be brought into play again. It was one more way to gain an understanding of what kind of rock this might be and how rewarding it could be as a test target.

"MAHLI has several roles when deciding about drilling," Edgett said. "One of them is characterizing the rock itself, as in, is this the stuff we want to drill? In this case at Yellowknife Bay, after they brushed the rocks, they cleared away the dust, the MAHLI could look at it and we could see that it was very fine-grained. So fine-grained that you really couldn't see the individual grains in MAHLI images, which told us the grains were smaller than fifty microns—how much smaller we really couldn't say. We don't have a microscope, but that was good enough to say 'We have a rock where there are no grains larger than fifty microns, so it is very well sorted and very fine.'" This is referring to the evenness and smallness of the little bits of sand that make up the rock, not how pretty it is . . . "That leads you down the path of something like a mudstone because of the grain size."

Between this diagnosis of mudstone and the hydrated minerals in the veins, Klein was looking like an ideal place for the first drilling attempt.

After taking some down time over the Christmas vacation, the crew was back and ready to drill in January. After a bit more reconnoitering, they performed pre-drilling tests on John Klein before boring into it.

The first step was to check pressure loading on the arm. The arm operators pressed the drill mechanism up against a few different parts of the rock, then measured the pressure feedback on the robotic arm to see if it was equivalent to expectations. It was within spec. This also allowed them to make sure that the area was stable enough to be drilled—you don't want your target tilting or skittering away or rolling while you are pounding and grinding on it with the drill bit. In Klein's case, the rock was a flat tablet of mudstone, so unexpected motions were not a concern. But in future drill attempts, the motions of more complexly shaped rocks certainly would be, so it was considered a good idea to establish proper procedures at the start.

The arm operators pushed the drill gently against Klein and left it there overnight to check out how much the thermal contraction that I mentioned earlier would affect things. It was possible that once you took the structure of the rover and arm into consideration, the whole thing could shrink as much as a tenth of an inch. Fortunately, for this kind of test, without a drill stuck in the rock, the most likely outcome was for things to pull away (contract), not push (expand)—the arm would contract up and back, simply pulling the drill head out of contact. When the rover and arm did expand in the morning hours, it would be returning roughly to the position it began in the day before. Dan Limonadi explained: "We don't plan on leaving the drill in a rock overnight once we start drilling, but in case that happens, it is important to know what to expect in terms of stress on the hardware . . . this test is done at lower preload values than we plan to use during drilling, to let us learn about the temperature effects without putting the hardware at risk."

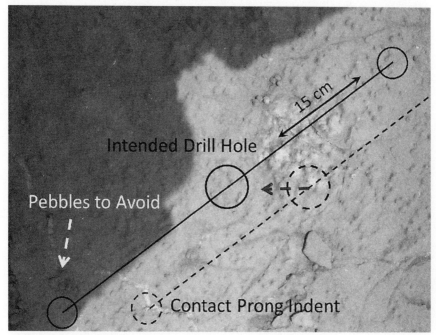

Fig. 28.1. PLANNING: Drill placement was planned to the last detail on an image from MAHLI. Here you can see the drill hole's planned position (the circle at center) and the contact points of the stabilizing posts, or prongs (the two circles to the lower left and upper right). Above the lower left indicator are five tiny pebbles that were marked off-limits, as "Pebbles to Avoid." *Image from NASA/JPL-Caltech/MSSS.*

There were a few other considerations when picking the target sample. Though it is unlikely, there are some minerals in Gale Crater that might gunk-up the drill. Remember that the way the drill gathers a sample is not by coring, but by grinding some rock dust, walking it up the screw cutouts on the drill bit, and collecting it in chambers for sorting and sieving. This is all predicated on a *dry sample*—if it is wet, or acted wet or syrupy, it might clog the sample system, possibly permanently. Even dry minerals can *act* like wet ones under certain circumstances.

The good news was that the vast majority of Mars is a dry desert and Gale appeared to be arid. Also, the rock in question did not appear to be of a type that, once ground, might behave poorly. The less good news: many of the rocks they wanted to look at, and possibly drill, were clays. Though these are outwardly dry, they can release water when drilled, and that would be risky. So there were many variables, and one always seemed to be tugging at the other.

A note: While this was the first time anyone had drilled on Mars, other celestial bodies had been drilled in the past. The US Apollo-program astronauts took many core samples on the moon, some six feet deep. The Soviet Union's Lunakhod lunar rover did the same thing, and the Soviets even took some shallow cores on Venus. But this was a first for Mars and also for the US planetary-exploration program. A lot was riding on this first drilling event. Planners hoped to do at least nineteen more drilled samples after this one, and hopefully well more than that. Without any experience drilling on Mars, patience and caution were the watchwords.

So, over the next few days, they conducted *more* tests—slowly and carefully. They applied more force to the rock and arm. They percussed the drill just a bit against the rock to ensure that the data looked right, and that loads were within expectations. They made a small, tentative predrill, which brought up just enough powder—onto the surface, not into the CHIMRA—to ensure that the consistency of the powder was what they wanted. This also ran the drill in both twisting and hammering motions, ensuring that it functioned as hoped when in contact with rock.

Brian Cooper, the rover driver, recalls some of the stress he was experiencing: "We did take baby steps. Drilling [is] not something we do quickly because we are not able to do it very often. There's a whole crew of people whose entire focus is on the drilling campaign. We check for slip on the rover, slipping by the wheels, every day. We are very sensitive to that, and we've never seen any noticeable slip due to thermal cycling [heating and cooling of the rover and arm]. But we would always keep that in mind, especially when you have this drill in a rock and you have a thermal cycle, you're getting a lot of stress on it. So drilling is planned as a single-day operation, but you might be stuck in a hole due to a fault, or something happens to the rover . . . you have to assume that might happen; you have to program it so that you wouldn't [have this] happen [and to ensure that] the arm would be safe in that situation."

It was ten sols between the beginning of these tests and the actual drilling—and they had already been at the drill site for over a month. That's an abundance of caution. I suspect there was at least one engineer who wanted just one more test of *something*, but the vote was in: time to go ahead.

The night before the big event, I had a chance to speak to Vandi Tompkins again, who was intimately involved with the drilling preparations, and she summarized the previous few days. "We'll be drilling this weekend! We have simulated the actual duration of the activity, and then there's all the time you spent coming up with the simulation. We have simulated and reviewed the entire operation." She seemed very comfortable with the thoroughness of the preparations. "For drilling not only do we have to place the arm exactly, but we have to do a whole set of things. You take a MAHLI image . . . then you place the APXS instrument so they have the initial assessment, because drilling is investing a lot of resources, so we tackle those other measurements first, make sure it's worth it. And then you place the arm in position just to preload it, you put pressure against it, so when you're drilling it won't jerk and slide away. All those things take time—anywhere from ten minutes to a couple of hours. Of course, the drilling itself takes time, too. The arm is very busy, not just for the MAHLI Mosaic [collection of images], but once you are sample processing, that can take hours of moving the arm around, getting it in the right place, when compared to MER, the arm motions are whole lot more complicated and [take] much longer."

It was February 8, or sol 180, and Curiosity was ready to drill at last; it would commence the historic operation tomorrow. But first, another long, cold night.

CHAPTER 29

RED MARS, GRAY MARS

February 9, 2013. Sol 181. Whatever you wanted to call it, this was the big day. The drill was in position, and the sequence had begun. Midspeed on the drilling, level 4 of 6 on percussion. The drill entered John Klein like a dentist working on a rotten tooth.

Remember that this is a semiautonomous operation, and there was still a substantial delay between Earth and Mars as they tracked the progress of the drill. And with all those parameters to worry about . . .

"We have accelerometers and an inertial measurement unit which monitor the pitch of the vehicle," said Limonadi. "We also have a force sensor in the arm, and we have the engineering cameras that can take pictures of the terrain and then take a picture again five minutes later to see if the rover moved. So we are effectively using onboard sequences when we do contact science. We turn on sensors on board and if any of those three things, the accelerometers, the arm force sensor, or the visual geometry, change; if any those things indicate that something's changed, that would stop the operation." He smiled tiredly. This may just be a bit more stressful than his search-and-rescue work for the county. "We just haven't done this before with the drill. It's a little bit like a mini EDL for us. My job as a project manager is to not break the sample system." He's pretty serious about the last part.

They didn't break the drill. In fact, it ground through the rock much quicker than anticipated, the stone was so soft. The resulting hole was a bit smaller in diameter than a dime and about 2.5 inches deep. As expected, the "tailings," the powder from the ground and beaten rock, traveled up into the drill's collection chambers. That powder needed to be sorted and sieved, which was how the grains—dust, almost—would be reduced down to the 150-micron size the instruments needed. That meant delivering the sample to the scoop, and from there, to CHIMRA, where the sorting mechanism and vibrator were.

Fig. 29.1. NO CUMBERLAND GAP: No gap here, the drill snugs right up to Cumberland, its second drill target. Drilling at John Klein, just a few feet away, looked about the same. Note the two metal posts on either side of the drill bit that make contact first to stabilize the turret and arm during drilling. *Image from NASA/ JPL-Caltech/MSSS.*

But the geologists were in for a surprise when they pulled the drill out of the hole. Since every part of the mission is recorded visually when possible, the sample itself was imaged.

Ashwin Vasavada warmed to the memory: "Seeing the sample was the best moment for me, coupled with the science team's realization that it wasn't oxidized." He grinned ear to ear as he remembered the day. "Mars before was red and now it's gray!" The contrast was even clear right there on the sampling scoop. "There was still some of the sample from Rocknest, some sand clinging to the back of the scoop. So you can see this bright red material we had sampled at Rocknest and then, right next to it, the gray material from John Klein. We said to ourselves, 'It's different!' That's why we have a drill to get inside these rocks where the oxidation wasn't there. That's the whole reason we did it, and it worked."

John Grotzinger remembered it as a high point, but one tempered with the patience he counsels; this is a story that is still unfolding: "When the picture came down and we got a look at it, I sent a message to the entire team and I said, "Look guys, Mars is gray!" To those of us who work on Earth [geologists], when you see a rock that you know has iron in it, as measured by the APXS, and now you've created a powder that's gray, there's a good chance that the iron is possibly not in a completely oxidized state. That's important because if there's fluid and if there has been organics, this is the kind of place that might be as good or bad as we're going to get for where the organics might be."

He continued: "We reached the total depth in six minutes. That's like a knife through hot butter when you are talking about rocks." You have probably never tried to drill into a rock, but even with this high-tech drill, it should have taken longer . . . "so we knew it was 'gray Mars' and we knew that it was soft. If you put both of those together, everything about it says mudstone."

To most of us, mudstone would be just that: hardened mud. But to a geologist, it implies so much more. As Grotzinger said, "where there is mudstone, there could well have been a lake and a habitable environment." Bingo: the bills for the MSL mission had just been paid with this single successful drilling. Gray rocks and soil equal a whole different Mars way back when.

"When I saw that and saw those data sets, I thought 'as far as this mission goes, we are golden. We've hit the jackpot.'" Grotzinger summarized. "We really couldn't ask for much more. We can turn the rover around and drive to the mountain to find the hydrated minerals, but everything that we had just found—right here—looks spectacularly good. The one exception is that we don't know if any of the minerals are hydrated. So we are not going to know until we get that powder into CheMin."

So off the powder went for analysis. Not only was the sample hydrated (that is, containing water molecules), but the surrounding region, as explored over the next couple of weeks, was as well. They had, quite literally, hit pay dirt. Gray pay dirt.

But I am jumping ahead a bit. Within three days, when bits of sample powder had been delivered to both SAM and CheMin, Curiosity's mission was back in prime-time status with the public, really for the first time (other than the "Martian Mystery" debacle) since landing. Then, right in the middle of the analyses, a problem; Curiosity went into "safe mode." It's very much what it sounds like, a bit like when Windows boots into safe mode after it detects a problem.

The problem was noticed on the afternoon and evening of February 27, when the rover failed to report in, uplink data, and go into its overnight sleep mode. The engineers switched from computer bank A to bank B again. NASA sent out a pun-drenched tweet on Curiosity's feed on @MarsCuriosity: "Don't flip out: I just flipped over to my B-side computer while the team looks into an A-side memory issue." It could have been advice to themselves—computer glitches are always a cause of deep concern.

It was those pesky high-energy particles, radiation from space, that constantly pour onto the Martian surface. Even with the radiation-hardened computers, occasionally a stray particle would smack into a 0 or a 1 in the binary-coded memory and cause corruption, and that's all it took to make Curiosity take a break to figure things out.

By March 4, they had things back on track. "We are making good progress in the recovery," Richard Cook said in a NASA press release. "One path of progress is evaluating the A-side with intent to recover it as a backup. Also, we need to go through a series of steps with the B-side, such as informing the computer about the state of the rover—the position of the arm, the position of the mast, that kind of information."

As the troubleshooting continued for the next two weeks, on March 16, another hiccup occurred: now the B-side computer was having memory fits. It too went into safety mode, leaving no confirmed backup computer. This said, the A-side had been fixed and was almost completely certified to go back into operation, but it was an uncomfortably close brush. If both went out at the same time, things would have been more challenging, or *interesting*, to say the least.

While all this was going on, there were still other science duties to attend to. The science team wanted a close look at that wonderful little drilled hole that had provided so much excitement. They turned to MAHLI.

"With the drilling, of course, we can only see down to the bottom that is visible," Edgett explained. "It turns out that when you look down the hole, you see a bunch of dust down there, which is the tailings or cuttings from the drill." These deposits filled the borehole to about halfway of its total depth, but they could still see enough. "From the Mastcam and the ChemCam, you could see in [the] walls this sort of a lightning bolt, a light-toned, zigzagging thing. Also, knowing that this depth was about three centimeters, they were able to accurately target their laser with ChemCam, going down the wall." They would be able to test at least the half of the hole that was visible to MAHLI. "We were able to see what made up these white vein minerals in the hole, and how much was something else. They also wanted to look down the hole to see if there were any layers. If there were fine, thin layers of sediment, that is there we would have seen them. So we ended up looking at it with the MAHLI, obliquely from four directions, so you can map the entire circumference around the wall."

Fig. 29.2. FIRST DRILLING ON MARS: The drill hole at John Klein. Subsequent to drilling, the shaft was shot multiple times with ChemCam (note the row of laser-burned pits), and the entire area was photographed with the MAHLI instrument. The drill hole is just slightly smaller than a dime. *Image from NASA/JPL-Caltech/MSSS.*

While the engineers were wrestling with the computer issues, there was another headlight coming down the tunnel. This was sample aging: the fresher the material is when it is analyzed, the better. The balky computer was slowing progress on sample analysis, and in April, just a few weeks away, solar conjunction would be upon them, when Mars would pass behind the sun as seen from Earth. While limited commands and monitoring are achievable during this period, "best practices" dictate shutting down and taking a break. Any transmissions could be corrupted, and it's just not worth taking the chance unless there are no other options. They needed to get things back on track as soon as possible.

By March 19th, they had confirmation that the B-side was back in good form, and they could continue science operations inside the rover, even as they also continued to resolve the original issue on the A-side of the computer.

Even as the computer drama was playing out, however, the scientists had continued their work when they could, while the engineers labored over the computer. Just before mid-March I was up at JPL doing some interviews, and I ran into Rob

Manning. I, of course, asked what was new, and he smiled like a kid with the biggest secret this side of Christmas: "I can't say much right now, but we think we found what we came for." The hairs on the back of my neck stood up a bit. "Habitability or organics?" I pestered. He chuckled and chided, gently, "Wait and see," so I did. I already knew a bit from previous interviews, but this could be something big.

The next day came the announcement.

<u>PRESS RELEASE</u>
03.12.2013
Source: Jet Propulsion Laboratory

NASA Rover Finds Conditions
Once Suited for Ancient Life on Mars

PASADENA, Calif.—An analysis of a rock sample collected by NASA's Curiosity rover shows ancient Mars could have supported living microbes.

Scientists identified sulfur, nitrogen, hydrogen, oxygen, phosphorus and carbon—some of the key chemical ingredients for life—in the powder Curiosity drilled out of a sedimentary rock near an ancient stream bed in Gale Crater on the Red Planet last month.

"A fundamental question for this mission is whether Mars could have supported a habitable environment," said Michael Meyer, lead scientist for NASA's Mars Exploration Program at the agency's headquarters in Washington. "From what we know now, the answer is yes."

Clues to this habitable environment come from data returned by the rover's Sample Analysis at Mars (SAM) and Chemistry and Mineralogy (CheMin) instruments. The data indicate the Yellowknife Bay area the rover is exploring was the end of an ancient river system or an intermittently wet lake bed that could have provided chemical energy and other favorable conditions for microbes. The rock is made up of a fine-grained mudstone containing clay minerals, sulfate minerals and other chemicals. This ancient wet environment, unlike some others on Mars, was not harshly oxidizing, acidic or extremely salty.

The patch of bedrock where Curiosity drilled for its first sample lies in an ancient network of stream channels descending from the rim of Gale Crater. The bedrock also is fine-grained mudstone and shows evidence of multiple periods of wet conditions, including nodules and veins.

Curiosity's drill collected the sample at a site just a few hundred yards away from where the rover earlier found an ancient streambed in September 2012.

"Clay minerals make up at least 20 percent of the composition of this sample," said David Blake, principal investigator for the CheMin instrument at NASA's Ames Research Center in Moffett Field, Calif.

These clay minerals are a product of the reaction of relatively fresh water with igneous minerals, such as olivine, also present in the sediment. The reaction could have taken place within the sedimentary deposit, during transport of the sediment, or in the source region of the sediment. The presence of calcium sulfate along with the clay suggests the soil is neutral or mildly alkaline.

Scientists were surprised to find a mixture of oxidized, less-oxidized, and even nonoxidized chemicals, providing an energy gradient of the sort many microbes on Earth exploit to live. This partial oxidation was first hinted at when the drill cuttings were revealed to be gray rather than red. "The range of chemical ingredients we have identified in the sample is impressive, and it suggests pairings such as sulfates and sulfides that indicate a possible chemical energy source for microorganisms," said Paul Mahaffy, principal investigator of the SAM suite of instruments at NASA's Goddard Space Flight Center in Greenbelt, Md.

An additional drilled sample will be used to help confirm these results for several of the trace gases analyzed by the SAM instrument.

Fig. 29.3. ASHWIN VASAVADA: As one of two deputy project scientists for MSL, Vasavada spends most of his time overseeing the processes that get Curiosity through its scientific tasks, interacting with the 480 other scientists on the mission. He is highly prized as a public speaker for his ability to impart scientific concepts in a way that laypeople can understand. *Image from NASA/JPL-Caltech.*

> "We have characterized a very ancient, but strangely new 'gray Mars' where conditions once were favorable for life," said John Grotzinger, Mars Science Laboratory project scientist at the California Institute of Technology in Pasadena, Calif. "Curiosity is on a mission of discovery and exploration, and as a team we feel there are many more exciting discoveries ahead of us in the months and years to come."

Wow. So it was official, ancient Mars would have been habitable to life as we understand it. Again, there was an indication of water, and this time, nice, drinkable water. And whatever had caused the oxidation of all that oxygen once present in the atmosphere—the reason Mars's sand and rocks are red—had not been present in the past. That was the "gray Mars" part of the announcement.

Vasavada put it elegantly: "We delivered the CheMin examination and got the results, which was wonderful, but I think by the time we got through all the press conferences for John Klein, a lot of them involved words like *oxidation* and *reoxidates*, but the bottom line is this: Mars is red but the inside of the rock is not, and that's the evidence that we were looking for. So that's great."

It was wonderful, like a second Christmas and New Year's combined. But then the inevitable train locomotive—that headlight in the tunnel I mentioned—caught up with the mission: solar conjunction. The mission went mostly silent for weeks. While a steady, weak telemetry stream came back through the furious and static-laced interference of the sun, the computer people, who had just resolved their issues, fretted. From April 9 to April 26, two and a half long weeks, Curiosity would be alone.

CHAPTER 30

GO SOUTHWEST, YOUNG ROVER

Solar conjunction was a quieter time at JPL—Curiosity was being monitored but was performing nondemanding activities while not driving. The rover sat parked until the solar interference ended. It gave a lot of people time to catch up, and some of them only then realized just how all-consuming the mission had been. It also gave the science teams and the drivers time to plan activities for postconjunction, and to do so armed with all the data that they had accumulated since landing. In that way, it was a blessing.

During this time, results of atmospheric-sample analysis from SAM were released at a conference in Vienna, Austria. The conclusion was that Mars had lost most of an ancient, denser, and more oxygen-rich atmosphere long ago. Also measured over the months of operation were humidity (a first on Mars), temperature, and wind speed.

By the end of April, Mars had moved past the sun and into the clear, so normal operations could resume. A second drill site was selected to attempt validation of the findings from John Klein. The new site, nine feet from Klein, was called Cumberland. It was chosen in part for the same reasons as Klein—flat and safe, but also because it *looked* like Klein and was made up of the same ingredients—right down to the gypsum veining. Any cross-contamination of the sample should be minimized. It was a good choice to verify the observations from the first drill sample.

Cumberland did have a somewhat-different texture, though, dotted with small bumps on the surface. These were ancient concretions, which were apparently formed in water. The drill hole was sunk on May 9 and was about the same size and depth as before. This time, once the sample was in hand, the rover drove off, with analysis scheduled to take place as other observations and driving were undertaken. JPL was anxious to get started toward Mount Sharp, five miles distant. Of course, the arrival date is predicated upon what surprises or points of interest the rover comes across on its five-mile trek to the foothills. Increased autonomy would also be required to get longer drives completed each day without as much guidance from home.

Along the way, a few locations, called waypoints, were preidentified for examination. The first, an outcrop called Darwin, was again representative of flowing water. It can be a challenge to keep the public engaged when all you have is another announcement about water (*public yawn*) on Mars, but each site is a bit different and all add to the overall story. And, of course, water is a primary focus of the mission and keeps everyone involved highly engaged—how water accumulated, how it behaved, and where it went are a continually unraveling mystery. Areas like Darwin, which contain conglomerates—rocky remains of ancient riverbeds—are favorite areas for detailed exploration.

Dawn Sumner is one of the researchers who worked Darwin and spoke of her experiences in a NASA press release: "We examined pebbly sandstone deposited by water flowing over the surface, and veins or fractures in the rock," she said. "We know the veins are younger than the sandstone because they cut through it, but they appear to be filled with grains like the sandstone." As if to echo my previous point, "We want to understand the history of water in Gale Crater . . . did the water flow that deposited the pebbly sandstone at Waypoint 1 occur at about the same time as the water flow at Yellowknife Bay? If the same fluid flow produced the veins here and the veins at Yellowknife Bay, you would expect the veins to have the same composition. We see that the veins are different, so we know the history is complicated. We use these observations to piece together the long-term history."

In February 2013, an interesting announcement was sent out from JPL, but this was not related to rocks . . . it was about the air. The SAM instrument had analyzed an atmospheric sample, looking specifically at isotopes of argon, and the results were compared with earthbound studies of meteorites. It had been long suspected that some meteorites found on Earth were pieces of Mars, based on their chemical composition. The new SAM results provided strong evidence to confirm this for some meteorites and disprove suspected Martian origins of others. The result has been, in effect, a kind of natural sample-return from Mars. The samples are very old and have traveled through space for millions or billions of years, but they are pieces of Mars and, as such, are of extreme interest. It's a chance to peer into Mars surface bits with all the sophisticated power a fully equipped laboratory on Earth can bring to bear, and without the cost of a $4 billion sample-return mission (though such a mission, still a high-priority goal for NASA, is obviously an entirely different scientific animal).

By the end of 2013, Curiosity was driving farther and faster than before, and making great progress. Then: more issues.

The first concerned the computer. On November 8, the computer performed a "warm reset," which is analogous to restarting your cellphone as opposed to performing the complete factory reset. In this case, the software spotted something it didn't like and reset itself to an initial, prior state. While still performing work, the machine was effectively in safe mode again. Within two days, the error was resolved and the computer returned to a normal state. The software team thought they knew the culprit: new software had seen something in older software—a "catalog file"—it didn't like and stopped the show until things were figured out. This was exactly what it was supposed to do.

But less than two weeks later, another issue cropped up, this one potentially more permanent. The rover indicated a "soft-short" in its electrical system, which, as opposed to a "hard short" (the kind that trips your circuit breakers at home), caused a change in voltage levels (as opposed to a shutdown). While not crippling, it was troubling and unexpected. Within three days, they knew the cause: the RTG power source hanging off the back of the rover.

If radioisotopic thermoelectric generators have a drawback, besides scaring the antinuke crowd, it is small short-circuits—little electrical malfunctions. Voyager had endured a number of these over the years, as had the nuclear-powered Cassini Saturn probe, and it seemed to be endemic with RTGs. The electricity-producing thermocouples that surround the plutonium are not working in the kindest, gentlest environment and sometimes act up, which is apparently what happened on Curiosity. That said, the Voyagers are still voyaging over thirty-five years later and still have power, though at slowly reducing levels. On Curiosity this was certainly not a serious setback, but it warranted keeping an eye on.

As the year neared a close, a happy result came in from analysis of the drill sample taken in Cumberland (the rover had been carrying that sample around like a brown-bag lunch since May and had taken occasional tastes since). For the first time ever, the geologists had been able to confirm the age of a rock on Mars. It landed on the timeline at between 3.8 and 4.6 billion years old, which was about what was expected.

And that's a good thing, because it lets you know that your hypothetical models are on the right track. It was another example of what the onboard instrumentation was capable of, as it had measured the amount of argon in the rock, a result of potassium that changes to argon over a long time span. Almost like carbon-14 dating, except that we're counting in the billions of years. It requires incredibly precise measurements to do this kind of evaluation, and Curiosity was fulfilling its promise.

Then in late December, halfway through its second Earth year of exploration, the first real trouble beset the rover. What started small quickly became a larger concern. The designers had known that the rover's wheels would sustain some damage as it went about its mission—remember that it's a full ton of mass pressing down onto six wheels, albeit in reduced gravity. The wheels were beginning to show more wear than anticipated—and by January 2014, a *lot* more wear was visible. The damage had gone from dents and pinholes to full-on rips the size of your thumb.

The problem was the particular type of terrain the rover was crossing as it neared an area called Dingo Gap. The terrain had thousands of sharp rocks—not unusual for Mars—sticking up in an unfortunate way—they were welded to the ground. One theory opines that these rocks, which on average appear to range from about the size of a golf ball to the size of a baseball, were part of an overall rock formation that was worn away by eons of windblown sand, whittling some of the remaining material to dangerously sharp protrusions. Normally when Curiosity drives over such rocks, they might sink into the soil or shift a bit, reducing wear and tear. But, stuck in the ground as they were, the sharp rocks created what looked like a road-hazard test course just waiting to puncture a wheel.

The rover drivers soon had to slow the pace again, being careful to pick the best and safest route. The rover drivers would stop frequently and use the MAHLI camera on the arm to photograph the wheels, which was a time-consuming operation. And even then, the damage was accumulating faster than anyone wanted.

If you look at one of Curiosity's wheels close up, it's actually amazing how thin they are. About the size of a small beer keg, they are machined out of a solid piece of aluminum. The wheel has thick, zigzagging cleats or ridges machined into its surface. But the expanse of metal that separates these is surprisingly thin. As always, it was a trade-off between strength and weight. The wheels can sustain a lot of damage before failing, but MSL is intended to be a long mission and there are many miles of driving ahead. The punctures were certainly not making anyone happy.

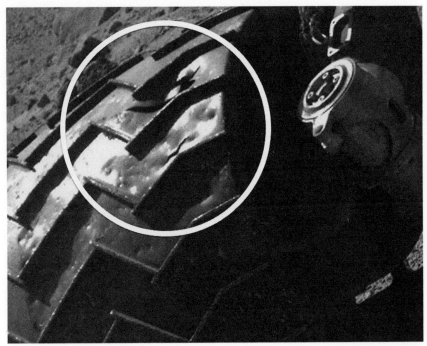

Fig. 30.1. SHARPIES: About sixteen months into its mission, Curiosity began crossing a flat area near Dingo Gap that had sharp rocks embedded in it. These wind-sculpted formations created the Martian equivalent of a series of spike strips. This image shows damage to one of Curiosity's wheels as the rover crosses the unforgiving terrain. *Image from NASA/JPL-Caltech/MSSS.*

At Dingo Gap, the drivers found a different route toward Mount Sharp that appeared to have fewer wheel-eating rocks. The only issue was that the rover would have to cross a sand dune to get there, and after the trials of the MER rover Spirit, which not only got stuck in sand a few times but ultimately died in a sand patch, the people charting Curiosity's path wanted to be very cautious in their approach.

By early February 2014, it was a *fait accompli*. JPL had decided that the risk was worth the reward, and Curiosity crossed the sand dune without mishap, then headed southwest into a depression and smoother terrain.

The road to Mount Sharp lay ahead, and the majestic foothills beckoned.

CHAPTER 31

DISCOVERIES!

S o, what is the sum total of discoveries and new information gathered by Curiosity since landing on Mars? Let's just say that it would fill a book far larger than this one. In fact, I have a book to my left right now, called *Mars Science Laboratory*, that was edited by John Grotzinger, Ashwin Vasavada, and a gent named Chris Russell. It is a collection of papers and chapters written *before* the landing, and is nonetheless over 850 pages long. If you can find a copy, it retails for $270.00. And no, my copy isn't for sale. Besides, there are dozens of new, postlanding papers that have been authored by the Curiosity team, with many more in the works. Each one is a colossal effort, involving a lot of time, cooperation, and, ultimately, peer review. It's a whole lot harder than writing a book like this one, and I don't envy them the task—but it's part of academic and scientific life. At the rate they are making discoveries, there is a lot of writing and peer-reviewing to be done.

So, let's rephrase our question: What are some of the *major* discoveries made by Curiosity and the science team since landing in August, 2012? They are many, and in a number of disciplines. In broad terms, there are geological, atmospheric, radiological, and chemical findings. Before we look at these, a quick rundown on Martian geological history might be in order.

As you are doubtless aware already, Mars formed when the other terrestrial planets did, at the birth of the solar system about 4.5 billion years ago, give or take a few tens of millions of years. The rocky planets formed inward of the asteroid belt, and the gas and ice giants formed outside it, beyond the so-called frost line.

Early on, there were lots of little protoplanets wandering around our neighborhood, blobs of planetary stuff that had not yet solidified into what they are today. For about 100 million years they smashed into one another, combining and destroying each other with abandon like a teenaged demolition derby. There are, of course, various theories about the specifics, but in general terms, the resulting mess coalesced into the four terrestrial planets we now see: Mercury, Venus, Earth, and Mars. A fifth that missed the train, never consolidating, now forms the asteroid belt beyond Mars. It is the wayward, shiftless cousin of the rocky planets.

For about 400 million years, things seemed to be finding a groove, but then came what is known as the Late Heavy Bombardment. There are a number of theories how it happened, but, in short, it did. One broadly acknowledged idea is that the gravitational forces of the large outer planets caused gravitational "resonances," which seem to have propelled wandering rocky material in toward our part of the solar system. The mass of these larger planets—Jupiter, Saturn, and the rest—caused this drifting rocky material to change orbit from roughly circular to an ellipse, and as they dipped into the inner solar system, some stayed there. It's worth mentioning that though the gas giants are big, they are not very heavy (in terms of mass) and that the sun still accounts for almost 99 percent of the total mass in the solar system, so our star is like a big gravitational magnet. This lasted several hundred million years more and resulted in the scarred planetary surfaces we see today, as these clumps of rock slammed into the inner planets. Interestingly, the end of this era seems to coincide generally with the beginnings of life on Earth.

On Mars, this era included and was followed by the geological epochs below:

PRE-NOACHIAN: About 4.5 to 4.1 billion years ago. This era encompasses the worst of the bombardments, and some of Mars's worst global scars came about during this time, notably an area called the Hellas basin and the differential elevations between the northern (basins) and southern (highlands) hemispheres.

NOACHIAN: About 4.5 to 3.6 billion years ago. The major craters and basins we see today were probably formed about this time. The Tharsis Bulge—a huge, three-thousand-mile-wide, four-mile-high lump on the Martian northern hemisphere—came into being. Mars was also deluged with water-bearing meteors and asteroids during this period, resulting in sufficiently large bodies of water to shape some of the larger fluvial (water-worn) features we still see today.

HESPERIAN: About 3.6 to 2.6 billion years ago. The extensive lava plains formed during this era, and possibly Olympus Mons, Mars's largest volcano. More water arrived as well, creating more fluvial features and possibly oceans. The later part of this era is also perceived as the general boundary between the warm, wet Mars and the cold, dry Mars we know today.

AMAZONIAN: About 2.6 billion years ago to the present. Fewer large meteors were slamming into the surface, but other activities continued. Lava flows and glaciers were active during this time. Mars was drying out and getting colder.

Note that these dates are estimates and especially the last boundary—Hesperian to Amazonian—could range from over 3 billion to 1.5 billion years ago.

The big news in terms of gross geological theory is that of sedimentation. Mars was long thought to be primarily volcanically formed and sculpted, with some wind and possibly water at work. But since the 1990s, it has become increasingly clear that a huge amount of interaction between water and the surface of Mars occurred, with the resulting sedimentary records included in its wake. Weather (with water) and not volcanism formed the face of the planet. Some of these formations are small and local, but they can also range up to hundreds of miles. The oldest of them go back over 4 billion years, to the later Noachian period. Nothing similar exists on Earth, with its more active (and comparatively poorly preserved) geological history. And Mars does not appear to have measureable tectonic activity, so what happens on Mars stays on Mars, so to speak. Also, without tectonics, there is not the uplift and rifting that occurs on Earth, so sedimentation seems to occur mostly within craters and water-worked channels. This is good for finding flat sedimentation, but it's not so good for seeing extreme or deep strata exposures, except for water-worn valleys and craters.

Catastrophic flooding, groundwater bursts, and even glaciers seem to be some of the bigger engines of water-caused change on Mars. The old Mars—warmer, wetter, and more active—caused weathering and change through these watery activities. As Mars lost its atmosphere and much of its moisture, the environment became dry, cold, and more chemically nasty. Wind and chemistry do more to erode what's left than anything else. The planet also oxidized as it cooled, resulting in the rust-red color we know today. The majority of the large water-formed canyons, and the resulting generation of sediment, seems to have occurred between the Noachian and the early Hesperian periods.

The largest strata belts—huge layers of sedimentation—appear to be concentrated in a roughly fifty-degree-north to fifty-degree-south equatorial band. This makes some sense, as this would be the more temperate part of Mars, and it would allow for vast amounts of rapidly flowing water. In places like Arabia Terra, such stratification can be seen in amazing relief—it looks like a topographic map brought to life, complete with bands of strata such as we are used to seeing on Earth. Whether wind or water were more responsible for this region's appearance is still under debate.

Most of the strata are more subtle than that seen in Arabia Terra and are only evident upon closer examination—like the heap of sedimentary rock at the center of Gale Crater. This eighteen-thousand-foot-high mound is one of the larger single exposures of continuous sedimentary strata on the planet, which made landing there a fine choice.

Mount Sharp, then, is like a time machine. At its base, Curiosity will be staring back almost 4 billion years into the past and, as it climbs the serpentine foothills, will slowly advance toward the modern era. But not all the fun must wait for the central peak to be reached.

Exploring Mars, or indeed any world, via robots requires a lot of planning and patience. But perhaps the most essential skill is inventiveness. So often, the ideas and theories held upon the inception of a Mars mission have changed radically by the time it is ready to fly. Then, concepts held dear at launch may have evolved by the time your machine lands. Finally, once on the surface, new discoveries force old ideas to adapt and mature. It's a constantly changing kaleidoscope of theory, observation, and problem solving.

One magnificent example of this is the search for habitable environments on Mars and how this primary element of Curiosity's mission has already shifted after less than two years. As John Grotzinger said it, "We have reached a turning point in this mission—it has changed from a mission to search for habitable environments to the search for organic carbon." Why would he say that so soon?

The answer lies in learning and adapting. When Curiosity drilled at John Klein and Cumberland, the geologists completed their primary analyses over the course of a couple of weeks. It became clear that these two drill holes, only nine feet apart and more flat and safe (for the rover) than remarkable in nature, held secrets we could only have dreamed of a decade ago. Remember red Mars/gray Mars? Those soil samples showed that the environment extant when those deposits were laid down in that wet, flowing stream were nearly perfect for life, at least of a microbial nature. The water was not salty. It was not too acidic or too alkaline. It was cold but not freezing. And it was filled with minerals in varying states of oxidation. These conditions are not only benign for life but provide enough energy and the proper nourishment for some types of microbes.

So we know that life *could* have existed . . . but how do we find out if it actually *did*? That's the 2.5-billion-dollar question.

Some form of organic molecules have been detected on the surface of Mars, as elaborated upon in the AGU press release. How much might have been a sampling-system contaminant from Earth, how much might have been meteoritic in origin, and how much might be biological versus abiotic (nonlife) in origin are all open questions. The key now is to test, retest, and test again. Then, validate results and look for overlaps and relationships.

Hold that thought for a moment—about the process of finding organic molecules, and possibly a biosignature of life. There is another huge complication to finding life on Mars—one that goes beyond the understanding of ancient environments. This complication is radiation.

Curiosity is the first mission to survey the radiation environment on the surface of Mars. This simple instrument, the RAD instrument, has been wonderfully successful at its task. But what it has told us is not a happy story. Mars is like your brain on drugs—one big, hot skillet. For although it is a cold place, the surface is hot with radiation.

A primary purpose for the RAD instrument was to measure how much radiation exposure astronauts would receive during a typical Mars mission—and it's a lot. About 1,000 millisieverts, or ten times what they would be bombarded with during a long mission on the International Space Station (ISS). But this exposure would be manageable if there were some shielding and protection during their voyage to Mars and their stay there.

But this same radiation, the result of high-energy particles streaming in from the sun and from deep space, has another downside. It may be actively frying any organic molecules in the surface soil and rocks of Mars. It is estimated that any area exposed for some tens of millions of years, as most are on this relatively unchanging world, would be sterile down to about ten feet below the surface. The drill can penetrate only a few inches. You can see the problem. A longer drill would be better, but such a device will have to wait for a later mission.

Add to this the chemistry in the soil—particularly the existence of toxic perchlorate—and you have a recipe for sterility. Moreover, perchlorate also causes organic molecules to burn up in the very tests designed to find them, as was discovered in relation to the Viking life-detection experiments. Drat.

But remember that I praised JPL's ingenuity? Read on.

The same team that compiled the RAD readings worked with the geologists to invent a way to skirt the issue of surface radiation bombardment and the havoc it wreaks on samples. In essence, let's say that you are on Earth and you want to find worms in your garden, what would you do? You could dig deep, where the soil is wetter and cooler, or you might simply turn over a rock to expose the damp soil, right? That soil would not have been baked out by the sun.

As it turns out, if you can calculate the rate at which the sedimentary outcrops in Gale Crater were being eroded, you can also calculate the amount of time the area previously "shaded" by the sedimentary overhang has been blasted by the sun and other radiation. For example, in a million years, wind alone would be enough to wear away about a yard of rock. So, if you drive the rover to the base of one of these eroded areas and snug right up to the edge of the part that is wearing away, what you are drilling into might have been exposed to radiation for only 500,000 or 900,000 years, instead of tens of millions. Which means—you guessed it— you wouldn't have to go as deep to possibly find organics because this surface has not been sitting under the blasting sun and deep-space radiation for as long as the "unshaded" areas. Of course, a half-million years is still a long time, but it's better than ten or twenty times that.

This sort of experiment is exactly what JPL sees ahead at its next major stop at Kimberley, which Curiosity reached in early April 2014. It is located about halfway to Mount Sharp. The rover will get closer to the site, snoop around, take some initial measurements, and then, if the scientists like what they see, pick a place to drill. If they can find the type of eroded overhand discussed above, they will drill there, into rock that has not been exposed to radiation for as long as the area surrounding it.

And now, let's look at some recent progress of the mission. In the latest round of announcements, which coincided with a slew of papers coming out in the journal

Science, the results of recent data from the rover were published. I'll go over some of the highlights:

- In a planetary exploration first, the geologists are figuring out how to date some of the rocks on Mars. Prior to this development, dating was done almost entirely by counting craters and guessing at their rate of creation, which is an inexact method at best. But, using new investigative methods that allow the science team to understand the rate at which potassium in rocks slowly converts to the gas argon, they are now able to accurately date rocks way, way back . . . we're talking 4.3 billion years old. There is a fudge factor of +/− 300–400 million years, but it's still a remarkable thing to achieve on a planet other than our own. And as it turns out, the results of this dating fits the old crater-counting methodology almost perfectly, validating that previous technique as a generally reliable tactic for determining general ages of broad areas of Mars. This means that there is far less guesswork involved in the dating of other regions that have so far been seen only by the Mars orbiters—that is, the vast majority of the planet. So it's a big deal. Expect some new theories about the evolution and nature of the planet's surface in the next few years.
- It is clear that there was both a source of carbon—CO_2—and nitrogen back when the habitat represented by the drill samples was extant. Between that and the benign nature of the ancient environment, life had a good chance of taking root. And the age of the sample was about 3.5 billion years, or about the same time that life first appeared on Earth. Coincidence? Nobody on MSL is speculating publicly—yet. But recall that there are plenty of smart people who posit that life may have formed on Mars first, and then migrated to Earth on meteorites. If evidence of past life on Mars, or even organic material with a biosignature, is found, then the timing of its formation will be critical. It may lead us toward an understanding of where life on Earth may have originated—and it may not be local to our own planet. It ain't over till the fat microbe (if any) sings.
- The lake bed in Gale Crater, the site of the drill samples, may have been as big as three by thirty miles in size, and it was probably much larger—similar to an upstate New York Finger Lake. Curiosity has given us the first con-

clusive evidence of a body of standing water that size on any planet besides Earth (this is excepting the possible oceans on Jupiter's moon Europa and Saturn's moon Enceledus—but these bodies of water, if extant, exist far below a layer of ice and are of a very different nature).

There is more, but those are the main points. Curiosity has already achieved one of its primary goals: detecting ancient habitable environments on Mars. Finding some organic molecules would be the biggest Christmas present anyone on the mission could ask for. Of course, nobody can predict that outcome. But, like the ten-year-old boy looking through the telescope at Griffith Observatory, I remain buoyantly optimistic.

Fig. 31.1. DUNKIN' DUNES: This image of the Shaler outcrop shows the cross-bedding the geologists get so excited about. An ancient flowing stream caused small sand dunes on the streambed, which eventually become these inclined layers. This area is thought to be younger than much of the rest of Yellowknife Bay. *Image from NASA/JPL-Caltech/MSSS.*

CHAPTER 32

A DRIVE TO MOUNT SHARP

When this book went to press, Curiosity had left the treacherous region that was smashing holes in its wheels and had taken a risky detour through Dingo Gap, crossing a thirty-two-foot sand dune, then driving across smoother terrain to reach the area named Kimberley. Currently, the rover has covered almost four miles. And, of all things, it is driving backward. Apparently this takes some of the strain off the wheels and would help to even out any more damage that might occur.

Fig. 32.1. THE ROAD TO MOUNT SHARP: From Dingo Gap, Curiosity sought a smoother route, one that would incur less damage to the wheels while still allowing for relatively swift progress. This image shows the pathway it took to continue the journey to its prime destination—Mount Sharp. *Image from NASA/JPL-Caltech/MSSS.*

On a recent autonomous drive, the rover crawled over three hundred feet in a single sol. That's a far stretch from earlier drives and a big vote of confidence in the onboard navigation. In the current terrain, there are far fewer sharp rocks, and those that do exist appear to be loose on the ground, meaning that they would shift and possibly sink a bit, causing less damage to the rover.

On the way to the mountain, Curiosity will of course be stopping at various waypoints or points of interest. That brings us back to its current location, and one where it will likely spend a fair amount of time, Kimberley. And what was Kimberley named after? Tick-tock . . . *wrong*, it was nothing in Canada. The earthly Kimberley is a spot of geological interest in Australia that contains very old rocks. And, as it turns out, both gold and about a third of the world's diamonds. On a side note, the only landing in Australia by the Japanese in World War II was at Kimberley. Four Japanese military officers spent about twenty-four hours there doing a bit of spying before wisely returning to their base. Isn't that interesting?

Kimberley (the Mars version) affords another chance to grab a drill sample, and at this time a couple of prime locations to do so have been identified. These provide the best chance yet for an overhang-protected (shaded) and younger sample. It is also one of the stops roughly bounding an area that is special for another reason.

"These are points of interest that help us piece together the geology between Yellowknife Bay and the lower slopes of Mount Sharp," says Ashwin Vasavada. "At some point the materials stop coming from the crater rims and start coming from Mount Sharp, and there's some very complex inner bedding that we can't figure out from orbit. It's probably also very difficult to figure out from the ground, but at least we can try. And so we've chosen four spots on the way to Mount Sharp that are almost like natural road cuts into the planes of these layers, places where wind probably eroded the hollows where we can see layers of materials exposed."

Melissa Rice added that this was a region the scientists had been coveting for over a year: "This is the spot on the map we've been headed for, on a little rise that gives us a great view for context imaging of the outcrops at Kimberley." Context imaging is the process of looking at areas near the one you are interested in to get a better idea of how the overall region was formed.

Then, after a few more stops, Curiosity will enter the foothills of Mount Sharp. As of now we can only look at orbital imagery and telescopic pictures from

Curiosity to try to get a sense of what kind of terrain the rover will encounter, but if what we see is any indication, it will be a grand adventure.

Fig. 32.2. PENULTIMATE DESTINATION? Curiosity's destination is these rugged foothills at the base of Mount Sharp. The sedimentary layering should keep the rover busy for a year or more. I did not say "final destination" because if all goes well with Curiosity and its nuclear power supply, it should live long enough to cross through the foothills and continue its adventures beyond, perhaps for many years. *Image from NASA/JPL-Caltech/MSSS.*

The many stops between Bradbury Landing and Mount Sharp, including Yellowknife Bay and now Kimberley, have pleased Grotzinger. They have also served to validate the contentious debate about landing sites for the mission. He can remember when, during that process, "There were a lot of people that didn't want Gale Crater, that took the rather-pejorative view that [Mount Sharp] is just a pile of windblown dust. 'You guys are going to go in there and not find anything!' Now, that could still be the case. But the point is, that pile of windblown dust must have been altered with water because we see clays and we see the hydrating sulfates." And that alteration is, after all, a driving force of the mission.

Grotzinger continues: "The question is, are all these layers [in Mount Sharp] a result of wind blowing material around? Or are they a result of water transporting sediments and building layers? Are they a combination of both? What we got in Yellowknife Bay was five meters of stuff that looks like it was pretty much just transported by water, including the possibility of a standing body of water on the lake. But we don't know how the age of those rocks at Yellowknife Bay relate to the strata that make up Mount Sharp. One version is that maybe we have already been at the base of Mount Sharp." By this he means that what was found at Yellowknife might be very similar to what they will find at Mount Sharp. "Maybe if we could scrape all the dust and the stuff away, we would go look and see the layers at Yellowknife Bay, they just go right down and beneath Mount Sharp, that was the very oldest stuff. There's another version, that there's a big unconformity where layers were deposited, layers were eroded to leave the mound and then the crater rim had its evolution with the flowing water. Then we had the alluvial fan come down and [provide] the sediment that [filled the interior] of the crater. So it could be that the Yellowknife Bay is actually the youngest stuff we'll see there." So Mount Sharp might be a repeat, in geological terms, of what has already been seen, or it may represent material far, far older than they have examined. We will have to wait until Curiosity gets there to find out.

And how will the drive into the foothills be planned and accomplished? "I'm setting up a group right now called Mount Sharp Ascent Team," Grotzinger says. "As much as possible, I will coach some of them myself because it's so important. But the goal of this is to actually quantitatively figure out how the hell we are going to go way up in there." But they are well prepared. "We did enough work on it to know qualitatively that it is doable. We know the slopes well enough. We know from the surface process materials that even though there is windblown sand there, that if I take the scarecrow [Curiosity's earthbound double] out to the Mohave Desert that we can handle going up slopes of at least twenty degrees. We know that from the HIRISE images that we can work our way up without worrying about any slopes where we get too deep to drive." It looks like the coast is clear to continue into the foothills once the rover has completed its work at Kimberley and a few successive stops.

He concluded: "We have a lot to do [to understand] how to use the instruments on the way up there, the materials that we are going to get and how are we going to understand the ancient history of water. Are we going to be able to measure the isotopic composition not just of the water in the modern-day Martian atmosphere, but [of the] rocks and minerals that we are able to place in SAM and so forth? We are still figuring out how to do that." Then, with a tired smile, he added, "It is *never* boring. Absolutely never boring. There are times when you can catch me on a bad day, when I have [had] it up to here with MSL . . . but you will never hear me say it's *boring*." He goes on to remind me, and not for the first time, that while he is nominally the leader for Curiosity's study of Mars, the *team* makes the choices, writes the papers, and takes the hard knocks together . . . nearly five hundred of them. The message is clear: he is not an emperor, he is not out to seek the limelight. He is here merely to lead the team to consensus and to inspire their best work.

I asked him to contextualize the foothills of Mount Sharp in a way most of us could relate to. "Well, when we start driving into Mount Sharp we will return to our mission objective, which is to explore these foothills that are about the size of one- to three- [to] five-story-tall buildings with narrow canyons between them. That will be what's it like as we get to that terrain. It's going to be, we think, visually quite beautiful. But those layers that you see there . . . well, we know from orbital data that they have clays and sulfates. So we hope to be able to able sample a number of what could be different habitable environments." He sighed. "We have a long drive. We think that it would take . . . if we stop to smell the roses a little bit, it could take until the end of the operable mission. I know that NASA is going to back us and so we have confidence that we'll continue on with this great exploration mission."

Amen to that.

YOU GOTTA HAVE CURIOSITY

S o what revs up the people involved on this mission about the future? What do they dream about at night when they dream of Mars? Or, as Grotzinger puts it, what is their "dinosaur-bone moment" (imagine finding that on Mars . . . yeah!)? Here is what many of the lead scientists had to say.

John Grotzinger, Project Scientist: "In Yellowknife Bay where we drilled, we realized that the scent was getting warmer and warmer. We looked at Mars as it changed from red to gray. What we got was an understanding of a habitable environment which is not just another discovery of water. . . . [It characterizes the] water and rocks that were there to tell us that the simple microorganisms, so simple that they don't need sunlight, they need only chemical energy, could live there."

He continued, "This was a truly integrated set of observations—every instrument was involved in this. It all added up to understanding this environment as being chemically one that was favorable for life, [and] not in a harsh way but actually quite a benign environment that was very much like Earth."

Scott McLennon, Participating Scientist: "For me the next really big thing is getting on to Mount Sharp, especially now that we have the results from Yellowknife Bay. We found a habitable environment, but the age of it is younger than a lot of us thought it would be—it could be quite a bit younger. As we go into Mount Sharp, the whole paradigm for Mars-system evolution is this idea of a transformation from a relatively wet, benign setting to a much more arid, dry, and acidic environment. That transformation is supposed to be recorded in the layered rocks in Mount Sharp . . . to those of us who work in terrestrial geology, there's [a] sort of symmetry in the history of the two planets, and that's very exciting."

John Grotzinger again: "The story of MSL is that the increase in capability [of the rover and its instrumentation] leads to a scale of complexity that takes a toll on humans . . . it's tiring, it's exhausting. But we are all committed to it because, thank

God, the rocks delivered. . . . There's a ton of cool stuff that we can learn about the atmosphere of Mars, but at the end of the day, if it would have [gotten] down into Yellowknife Bay and found a bunch of unaltered basalts [vanilla Mars rocks] there, it would have been a tough moment for the mission. Because then what do we have to show for the first year of operations? As it turned out, we got really lucky and it all worked out. In fact, I wouldn't say it was pure luck, I would call it . . . serendipity."

Ken Edgett, Principal Investigator for MAHLI: "Think of the Martian [history] as a twenty-four-hour clock and right now, this moment, is just some tiny fraction microsecond before midnight of the next day. The rocks that we were looking at and exploring in Gale are back somewhere around in the 3 a.m. to 6 a.m. time frame of Mars history. It was a very long time ago, so Mars was a different place. At that time there were large craters forming more frequently than [there are] now. And there were volcanoes erupting, and all of those things produce sediments. The atmosphere was thicker, and wind transports sediment. We now think, and know, from all the missions over the last fifteen years that there was [also] water."

Edgett sees Mount Sharp as I described it—like a sort of time machine to go back to that ancient, earthlike Mars. "We can go there and see not only that there are habitable environments recorded there, but we can witness how these environments changed over time. As we go up the mountain we will see the time get closer and closer to now; what we see will be younger and younger."

David Blake, Principal Investigator of CheMin: "The next big thing that the community wants to see is the return of the Mars sample to be analyzed on Earth. That is a tremendous added value in itself, it is a tremendous undertaking. The Mars 2020 mission would take a Curiosity-class rover and land it on the surface and find habitable zones as Curiosity did, then collect drill cores that would then be placed in a returnable cache that would be collected by a second flight that would land and take that second cache of material and bring it back to Earth."

Ashwin Vasavada, Deputy Project Scientist: "About a year and a half ago, I was out . . . [filming an] outreach video about why we selected Gale as a landing site; and, walking around back there [behind JPL]—and it was springtime—water was coming down . . . the mountains and forming a little stream through the arroyo and carrying debris. . . . The stream was full of little pebbles, rounded as they make their way down the mountains, sand with them. . . . Looking around, I told the camera man as we were filming this video, 'if we ever found something like this on Mars, it [would] be a homerun.' . . . I think all of us . . . say those things, but in our hearts we knew it was a long shot, it wasn't to be taken for granted. And then you fast-forward to landing a year ago. As the rover landed, its rocket engines were scouring through the gravel. And the next morning, we get this picture from Mars and look in the scour mark—and, what do you know, its bedrock, [a] conglomerate of rock made of rounded pebbles and sand melted together. Even before Curiosity landed, she already hit this homerun for us and revealed this ancient Mars to us that [we] were hoping to find. So we drove literally through a streambed on Mars that flowed ankle-deep a few billion years ago. . . . I'm getting goose bumps just telling you that."

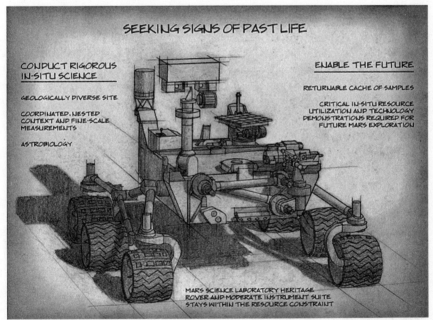

Fig. E.1. ONWARD! This depiction of NASA's Mars 2020 rover is tentative but shows the strong family resemblance to Curiosity. The instrument package had not yet been defined when this was released, but one idea currently popular is to have the rover find interesting samples, then package them up and leave them for a later sample-return mission to bring them back to Earth. Only then will we be able to really peel back the remaining veil of mystery surrounding Mars. *Image from NASA/ JPL-Caltech.*

Adam Steltzner: "Prior to landing, I'd been asking myself this question—why we do what we are doing, and why it is important to us? Although we go to Mars for the science questions, I don't think that those hundreds of people in Times Square watching the landing were there at 1:30 in the morning because they were dying to know about the pH, the salinity, and the environmental surface of Mars. I think when we explore, we're asking questions about ourselves as individuals, as a society, as a people. Neil Armstrong, I think, hinted at that with the words he chose to say when he set foot on the surface of the Moon: 'That's one small step for a man, one giant leap for mankind.' What he was hinting at was that he was carrying us with him in that exploration. I think Curiosity carries us with her when she's on the surface of Mars and helps us ask questions about who we are, how grand we are. What questions do we dare ask, and hope to be able to answer? I think [that] through it, we dream a little bigger, maybe aspire a little higher, and in some sense we're a little better."

Fig. E.2. SOMEDAY . . . The ultimate goal for the world's major space programs—those of the United States, Russia, and now China—is a crewed mission to Mars. It's been seriously discussed since the mid-twentieth century; it's time to step up the pace. Curiosity has provided data critical to protecting astronauts on the long trip to the red planet. *Image from Pat Rawlings/NASA.*

And perhaps Al Chen, the voice of EDL on that amazing night almost two years ago, said it best as he recalled something Steltzner said just before MSL entered the Martian atmosphere: "I really love our team. We did everything we planned to do and we were right on track. This is everything we could have wanted. As Adam said that night, 'We should do everything we can to deserve victory.' I thought that was pretty special. We may not always gain victory, but at the end of the day we should feel like we deserve it."

"We should do everything we can to deserve victory." . . . The people of JPL and their associated institutions did, and they continue to do so, every day.

Curiosity roves on.

SELECT BIBLIOGRAPHY

PRINT SOURCES

Baker, V. R., R. G. Strom, V. C. Gulick, et al. "Ancient Oceans, Ice Sheets and the Hydrologic Cycle on Mars." *Nature* 352 (1991): 589–94.

Barlow, Nadine G. *Mars: An Introduction to Its Interior, Surface and Atmosphere.* Cambridge: Cambridge University Press, 2008.

Carr, Michael H. *The Surface of Mars.* New Haven: Yale University Press, 1980.

DeHon, R. A. "Martian Lake Basins and Lacustrine Plains." *Earth, Moon, and Planets* 56 (1992): 95–122.

Edgett, Ken. "The Sedimentary Rocks of Sinus Meridiani: Five Key Observations from Data Acquired by the Mars Global Surveyor and Mars Odyssey Orbiters." *Mars* 1 (November 2, 2005): 5–58, doi: 10.1555/mars.2005.0002.

Ezell, Edward Clinton. *On Mars: Exploration of the Red Planet 1958–1978.* New York: Dover, 2009.

Greeley, Ron, and Paul D. Spudis. "Volcanism on Mars." *Rev. Geophysics Space Physics* 19 (2001): 13–41.

Grotzinger, John P. "Introduction to Special Issue: Habitability, Taphonomy, and the Search for Organic Carbon on Mars." *Science* 343, no. 6169 (January 24, 2014): 386–87, doi: 10.1126/science.1249944.

Grotzinger, John P., et al. "A Habitable Fluvio-Lacustrine Environment at Yellowknife Bay, Gale Crater, Mars." *Science* 343, no. 6169. Originally published January 2014; published online December 9, 2013. doi: 10.1126/science.1242777.

Grotzinger, John P., et al. "Mars Science Laboratory Mission and Science Investigation." *Space Sci Rev* 170 (2012): 5–56, doi: 10.1007/s11214-012-9892-2.

Grotzinger, John, Ashwin Vasavada, and Christopher Russell eds. *Mars Science Laboratory.* New York: Springer, 2013.

Grotzinger, J. P., J. Crisp, A. R. Vasavada, R. C. Anderson, C. J. Baker, R. Barry, D. F. Blake, P. Conrad, K. S. Edgett, B. Ferdowsi, R. Gellert, J. B. Gilbert, M. Golombek, J. Gómez-Elvira, D. M. Hassler, L. Jandura, M. Litvak, P. Mahaffy, J. Maki, M. Meyer, M. C. Malin, I. Mitrofanov, J. J. Simmonds, D. Vaniman, R. V. Welch, and R. C. Wiens. "Mars Science Laboratory Mission and Science Investigation." *Space Science Reviews* 170 (2012): 5.

Grotzinger, John, and Ralph Milliken, eds. *Sedimentary Geology of Mars.* Tulsa: Society for Sedimentary Geology, 2012.

Hassler, Donald, et al. "Mars' Surface Radiation Environment Measured with the Mars Science

Laboratory's Curiosity Rover." *Science* 343, no. 6169. Originally published January 2014; published online December 9, 2013. doi: 10.1126/science.1244797.

Hoyt, William Graves. *Lowell and Mars*. Tucson: University of Arizona Press, 1976.

Jacqué, Dave. "APS X-rays Reveal Secrets of Mars' Core." Argonne National Laboratory. Accessed March 2011.

Klein, H. P. "The Viking Biological Investigation: General Aspects." *J. Geophysics Res.*, 82 (1997): 4677–80.

Malin, Michael C., and Ken S. Edgett. "Evidence for Persistent Flow and Aqueous Sedimentation on Early Mars." *Science* 302 (November 13/December 12, 2003): 1931–34, doi: 10.1126/science.1090544.

Malin, Michael C., and Ken S. Edgett. "Sedimentary Rocks of Early Mars." *Science* 290 (December 8, 2000): 1927–37, doi: 10.1126/science.290.5498.1927.

Malin, M. C., K. S. Edgett, B. A. Cantor, M. A. Caplinger, G. E. Danielson, E. H. Jensen, M. A. Ravine, J. L. Sandoval, and K. D. Supulver. "An Overview of the 1985–2006 Mars Orbiter Camera Science Investigation." *Mars* 5 (January 6, 2010): 1–60, doi: 10.1555/mars.2010.0001.

McCurdy, Howard. *Inside NASA*. New York: Johns Hopkins University Press, 1994.

Murck, Barbara. *Geology*. New York: John Wiley and Sons, 2001.

Mutch, Tim A., Ray E. Arvidson, J. W. Head III, et al. *The Geology of Mars*. Princeton: Princeton University Press, 1976.

Neufeld, Michael. *Von Braun*. New York: Vintage Books, 2007.

Price, Pritchett. *The Mars Pathfinder Approach to Faster-Better-Cheaper*. Los Angeles: Pritchett Publishing, 1998.

Reiber, Duke B., ed. *The NASA Mars Conference*. San Diego: AAS Publications Office and Univelt, 1986.

Schorghofer, Norbert, Oded Aharonson, and Samar Khatiwala. "Slope Streaks on Mars: Correlations with Surface Properties and the Potential Role of Water." *Geophysical Research Letters* 29, no. 23 (2002): 41–44.

Wiens, Roger. *Red Rover: Inside the Story of Robotic Space Exploration, from Genesis to the Mars Rover Curiosity*. New York: Basic Books, 2013.

Zubrin, Robert. *The Case for Mars*. New York: Touchstone, 1996.

ONLINE SOURCES

BAE Systems. "RAD750 Radiation-Hardened PowerPC Microprocessor." July 1, 2008. http://www.baesystems.com/our-company-rzz/our-businesses/electronic-systems/es-product-sites/space-products-and-processing/processors. Accessed August 2011.

Dahya, N.; Jet Propulsion Lab, NASA, Pasadena, CA; and E. T. Roberts. "Design and Fabrication of the Cruise Stage Spacecraft for MSL." Aerospace Conference, 2008 IEEE. IEEE Explore. March 1–8, 2008. http://ieeexplore.ieee.org/xpl. Accessed March 2013.

National Aeronautics and Space Administration (NASA)/Jet Propulsion Laboratory (JPL)/ California Institute of Technology (Caltech). "Announced: Martian Diaries" (multiple entries). http://mars.jpl.nasa.gov/msl/news/whatsnew/index.cfm?y=2012&t=. Accessed August 2012–February 2014.

———. "Curiosity Capabilities: SAM." Goddard Space Flight Center. http://ssed.gsfc.nasa .gov/sam/curiosity.html. Accessed June 2013.

———. "Curiosity Makes Historic Landing at Gale Crater." http://www.nasaspaceflight.com/ 2012/08/msl-curiosity-historic-Martian-landing-at-gale-crater/. Accessed September 2013.

———. "Curiosity Rover Status" (multiple entries). http://mars.nasa.gov/msl/. Accessed August 2012–January 2014.

———. "Curiosity's SAM Finds Water and More in Surface Sample." http://www.nasa .gov/ content/goddard/curiositys-sam-instrument-finds-water-and-more-in-surface-sample/# .UwUpMkJdXes. Accessed January 2014.

———. "Entry, Descent and Landing Configuration." http://mars.jpl.nasa.gov/msl/mission/ spacecraft/edlconfig/. Accessed September 2013.

———. "Final Minutes of Curiosity's Arrival at Mars." NASA/JPL. http://www.nasa.gov/ mission_pages/msl/multimedia/gallery/pia13282.html. Accessed April 2011.

———. "Getting to Mars One Simulation at a Time." http://www.nasa.gov/centers/langley/ news/researchernews/rn_MSLEDL.html. Accessed December 2013.

———. "JPL Blogs: Dan Limonadi." http://blogs.jpl.nasa.gov/tag/dan-limonadi/. Accessed September 2013.

———. "JPL Blogs: Rob Manning." http://blogs.jpl.nasa.gov/tag/rob-manning/. Accessed December 2013.

———. "JPL MSL Press Releases" (multiple entries). http://marsprogram.jpl.nasa.gov/msl/ newsroom/pressreleases/20081119a.html. Accessed August 2012–January 2014.

———. "Mars Science Laboratory's Dynamic Albedo of Neutrons (DAN)." http://mars.jpl .nasa.gov/msl/mission/instruments/radiationdetectors/dan/. Accessed March 2013.

———. "MSL Landing Site Selection." http://msl-scicorner.jpl.nasa.gov/landingsite selection/. Accessed June 2013.

———. "MSL Mission Report." http://mars.jpl.nasa.gov/msl/mission/technology/insitu exploration/edl/skycrane/. Accessed November 2013.

———. "MSL Science Corner: CheMin." http://msl-scicorner.jpl.nasa.gov/Instruments/ CheMin/. Accessed September 2013.

———. "MSL Science Team Publications List." http://mars.jpl.nasa.gov/files/mep/msl_sci _team_key_papers.pdf. Accessed September 2013.

———. "NASA at Fall 2013 AGU Conference." http://www.nasaspaceflight.com/2012/08/ msl-curiosity-historic-Martian-landing-at-gale-crater/. Accessed December 2013.

———. "NASA at 2012 AGU Conference." http://www.nasa.gov/topics/earth/agu/nasa -agu-briefings-2012.html. Accessed December 2012.

———. "New Curiosity Research Papers" (multiple entries). http://mars.nasa.gov/msl/ mission/science/researchpapers/. Accessed September 2012–January 2014.

———. "Press Conferences at Fall AGU Conference." https://fallmeeting.agu.org/2013/ media-center/press-conferences/. Accessed December 2013.

———. "SAM Instrument Suite." http://mars.nasa.gov/msl/mission/instruments/ spectrometers/sam/. Accessed December 2013.

Wall, Mike. "Touchdown! Huge NASA Rover Lands on Mars." Space.com. http://www.space .com/16932-mars-rover-curiosity-landing-success.html. Accessed December 2012.

Webster, Guy. "Geometry Drives Selection Date for 2011 Mars Launch." NASA/JPL-Caltech. http://www.jpl.nasa.gov/news/news.php?release=2010-171. Accessed December 2013.

INTERVIEWS

Al Chen, May 3, 2013; August 6, 2013

Ashwin Vasavada, November 13, 2013

Bobak Ferdowsi, March 24, 2014

Brian Cooper, October 10, 2013

Dan Limonadi, December 18, 2013

David Oh, August 21, 2012

Doug Ming, December 18, 2013

Guy Webster, March 18, 2012; July 1, 2013

Jakob van Zyl, April 24, 2013

John Beck-Hoffman, June 12, 2012

John Casani, April 30, 2013

John Grotzinger, May 3, 2013; August 14, 2013; August 30, 2013; November 6, 2013; November 14, 2013; November 18, 2013; December 4, 2013

Joy Crisp, December 2, 2013

Justin Maki, December 12, 2013

Ken Edgett, January 23, 2014

Lauren DeFlores, November 12, 2013

Lawren Markle, June 12, 2012

Melissa Rice, May 1, 2013

Mike Malin, June 12, 2012

Mike Wall, June 12, 2012

Rebecca Williams, December 17, 2013

Rob Manning, February 10, 2012; August 12, 2012; August 4, 2013; October 22, 2013

Scott McLennon, December 12, 2013
Steve Squyres, June 27, 2011
Suzanne Dodd, August 22, 2013
Vandi Tompkins, May 16, 2013

INDEX

*Page numbers in **bold** indicate images within the text. See also the photo insert.*